Rafael Narciso Meirelles
Luiza R. Redaelli

Diachasmimorpha longicaudata

Rafael Narciso Meirelles
Luiza R. Redaelli

# Diachasmimorpha longicaudata

## From the laboratory to the field: releases in Rio Grande do Sul

ScienciaScripts

Cover image: www.ingimage.com

This book is a translation from the original published under ISBN 978-3-330-75484-3.

Publisher:
Sciencia Scripts
is a trademark of
Dodo Books Indian Ocean Ltd. and OmniScriptum S.R.L publishing group

120 High Road, East Finchley, London, N2 9ED, United Kingdom
Str. Armeneasca 28/1, office 1, Chisinau MD-2012, Republic of Moldova, Europe
Printed at: see last page
**ISBN: 978-620-8-26862-6**

# SUMMARY

*"An increasing number of people believed that it had been a terrible mistake for the species to come down from the trees. Some said that even climbing the trees had been a bad idea, and that no one should ever have left the sea. "*

Douglas Adams

## THANKS

I would like to thank Dr. Luiza R. Redaelli for her guidance during my 12 years at the Federal University of Rio Grande do Sul. Thank you to Dr. Simone M. Jahnke, who corrected and revised the manuscript that led to this book countless times and opened the gates to her backyard so that I could use her orchard to carry out the pilot tests.

To Dr. Miguel Francisco de Souza Filho and Dr. Valmir Antônio Costa, from the Biological Institute of São Paulo, for their help in identifying the parasitoids.

I would like to thank my colleagues in the laboratory who worked with me during the years in which this work took place, and whom I will not be able to name, so as not to take up too much time. Many have passed on and many have left beautiful marks, forming bonds of friendship. But two colleagues deserve special thanks, as they were the scholarship holders who worked with me on the project: Clàudia Ourique and Dania Ozorio. They were always on hand, learned a lot and I saw them grow as researchers. I'm glad to have been a part of it.

Now my family.

To my mother, a companion for everything in our family's life and, posthumously, to my father, who passed away in 2012, leaving behind, before that, the education that shaped my character. Together, my parents endured the hardships of life to keep their children studying. Now we will reap the rewards. They are the mainstays of my life, points of reference that serve as an example of character and love for their children.

I would like to thank my brothers, Màrcio, Marcelo and Lucas, and my sister Maira. They have always been willing to help or to give us the strength we need in the form of a compliment or their company.

Last but not least, I'd like to thank Paola Ramos for her love. It was during this work that we met and now we're at another stage in our lives. There's nothing like eternalizing my thanks in this book by saying that she has been very important to me. You've stood by me through the wonderful times and the difficult ones. She used to be a girlfriend, now she's family. And along with her, I thank her family, brothers and parents, who made me feel like one of them and offered me advice (father-in-law and mother-in-law) and a skilled travel guide (mother-in-law).

To those I've forgotten, but who feel deserving of my thanks: thank you.

# 1 INTRODUCTION

Fruit flies (Diptera: Tephritidae) are among the main pests of fruit growing in Brazil and in Rio Grande do Sul (RS). Both the damage caused by the development of the larvae and by the females ovipositing or testing the substrate cause losses that can reach almost the entire production.

In Rio Grande do Sul, the most abundant tephritid species are *Anastrepha fraterculus* (Wiedemann) and *Ceratitis capitata* (Wiedemann). The former is native and has been recorded in native and exotic fruit trees, while the latter, being exotic, is generally present in introduced species.

Control is often carried out using insecticides, which can contaminate the environment and leave residues on the fruit. This method of population reduction has become increasingly limited due to the demands of consumers in the domestic and foreign markets and the reduction in the number of products registered and permitted in fruit growing. In this context, pest control by natural enemies is an alternative to chemical control.

Natural biological control by native parasitoids of fruit flies does occur, but the parasitism rates recorded in RS are very low, rarely reaching values higher than 15%. Thus, the introduction of an exotic species could serve as a tool to increase these rates.

In classic biological control programs for fruit flies, parasitoids from the Braconidae family have been used because of their specificity and efficiency, with *Diachasmimorpha longicaudata* (Ashmead) being the most widely used species in the Americas.

*Diachasmimorpha longicaudata* was first used in Hawaii in the 1940s and later in Mexico, other states in the United States, Guatemala, Costa Rica, Panama and Argentina.

In Brazil, this parasitoid was introduced in 1994 and released between 1995 and 1998 in the Recôncavo Baiano, with the aim of studying its relationship with the native parasitoid guild. Since then, work has been carried out in São Paulo, Bahia and Minas Gerais.

Little is known about the possibilities of controlling fruit flies with *D. longicaudata* in RS. The objectives of this work were therefore: (1) to evaluate the parasitism rates of *D. longicaudata* and infestation by native species of fruit flies on loquats, pitangas, peaches, guabirobas, araçâs, persimmons, guavas and guava-serranas brought in from the field; (2) to observe parasitism rates of *D. longicaudata* in semi-field conditions on the same species of fruit *and* on kumquats. *longicaudata* in these same species of fruit and in *kumquats*, but artificially infested, and (3) to record parasitism before, during and after the release of *D. longicaudata* in loquat, peach, araçá and caquí orchards.

# 2. LITERATURE REVIEW

## 2.1 Fruit flies

### 2.1.1 Occurrence in Brazil and Rio Grande do Sul

Fruit fly species of economic importance in Brazil are distributed in just four genera of Tephritidae: *Ceratitis* MacLeay, *Rhagoletis* Loew, *Bactrocera* Macquart and *Anastrepha* Schiner. However, *Ceratitis* and *Anastrepha* are considered the most important, having a wide distribution in Brazil, while *Bactrocera* is restricted to the far north, represented only by *Bactrocera dorsalis* (Hendel). *Rhagoletis* has only four species associated with solanaceae in the Southeast (Oliveira & Paula-Moraes, 2006).

The *Ceratitis* genus in Brazil is represented by only one species, the Mediterranean fruit fly, *Ceratitis capitata* (Wiedemann), which occurs more frequently on exotic fruits (Carvalho *et al.*, 1998; Zucchi, 2000). The fruit families with the highest number of species related to *C. capitata* are Myrtaceae (21), Rutaceae (14) and Rosaceae (9) (Zucchi, 2000; 2008).

The genus *Anastrepha* is distributed throughout Latin America and 115 species have been recorded in recent surveys, found especially on native fruit (Zucchi, 2000; 2008). *Anastrepha fraterculus* (Wiedemann) is the species with the most associated hosts, with 109, including native and exotic fruit trees (Zucchi, 2008). According to the same author, the plant families with the most associated species of *Anastrepha* flies are Sapotaceae (19), Myrtaceae (16), Anacardinaceae (11) and Passifloraceae (9).

In Rio Grande do Sul (RS), the South American fruit fly, *A. fraterculus* and the Mediterranean fruit fly are the main species that cause damage to cultivated fruit trees, the former being considered a key pest and of greater importance (Kovaleski *et al.,* 2000). In citrus groves in the municipalities of Montenegro and Pareci Novo, *C. capitata, Anastrepha dissimilis* Stone, *Anastrepha grandis* (Macquart), *Anastrepha pseudoparallela* (Loew) and *A. fraterculus* were captured, the latter being the most abundant, accounting for more than 90% of captures (Silva *et al.*, 2006). According to Nunes *et al.* (2012), of the 13 fruit species evaluated in the municipalities of Pelotas and Capâo do Leâo, in the south of RS, only the coconut tree did not recover specimens of *A. fraterculus*, and only *C. capitata* was recorded. In the other fruit trees, the South American fruit fly was the most abundant. However, according to Dias *et al.* (2013), in the western region of the state of Rio Grande do Sul, close to the borders with Uruguay and Argentina, of the 19 fruit species sampled, *C. capitata* was the most abundant in ten, while *A. fraterculus* was the most abundant in only two fruit species.

In addition to individuals of Tephritidae, there are reports of flies from the Lonchaeidae family being caught on fruit collected in RS (Silva *et al.*, 2006; Gattelli *et al.*, 2008; Nunes *et al.*, 2012; Dias *et al.*, 2013). With regard to the insects in this family, there is disagreement as to their ability to cause damage. While some authors defend the idea that there are species that can become primary pests, causing damage to healthy fruit (Uchôa-Fernandes *et al.*, 2003; Raga *et al.*, 2004; Strikis & Prado, 2005), others suggest that lonqueids oviposit only in holes left by tephritids or in wounds in the skin of the fruit (Souza *et al.*, 1983; Malavasi *et al.*, 1994; Araujo & Zucchi, 2002a).

In RS, lonqueideos have been recorded on araçà, guiaba-serrana, guava, sky orange, Murcott tangor, carambola, papaya, pitanga, loquat, peach and bush cherry (Silva *et al.*, 2006; Gattelli *et al.*, 2008; Nunes *et al.*, 2012; Dias *et al.*, 2013).

### 2.1.2 Damage caused by fruit flies

Fruit fly damage can be caused by the larvae which, when feeding, open galleries inside the fruit, destroying the flesh (Aguiar-Menezes *et al.*, 2004), causing physiological changes, accelerating ripening and leading to premature drop (Salles & Rech, 1999; Botton *et al.*, 2005). The females also cause damage because, when they insert the ovipositor through the epidermis to lay their eggs, they leave wounds in the skin and flesh of the fruit. These wounds can lead to the entry of opportunistic microorganisms responsible for rotting and premature fruit drop (Aguiar-Menezes *et al.*, 2004; Santos *et al.*, 2008). In fruit such as pears and apples, the female fruit fly can leave a mark on the epidermis at the puncture site, causing deformation and depreciating their commercial value (Salles, 1995; Nora & Hickel, 2006). According to the author, this damage is irreversible and can occur when the female punctures the epidermis of the fruit just to test the substrate, leaving no eggs or when they are unviable. All these factors, as well as restricting exports due to quarantine barriers imposed by many countries, also make *fresh* trade unviable (Malavasi *et al.*, 1994; Salles, 1995; Brasil, 1999; Nascimento & Carvalho, 2000; Aguiar-Menezes *et al.*, 2004; Carvalho, 2005a).

### 2.1.3 Fruit fly monitoring and control in Brazil

Monitoring fruit flies should be considered an important stage prior to control, as this makes it possible to know the species and the abundance in which they occur in cultivation or study areas (Carvalho, 2005a). Fly traps or McPhail-type traps baited with a food attractant are the most commonly used for monitoring. The way this apparatus works is inherent to its shape and the attractant, as the fly tends to enter the jar without being able to get out, drowning in the liquid (Carvalho, 2005a).

In Rio Grande do Sul, fly-trap monitoring is carried out using fruit juices or formulated commercial products (such as hydrolyzed protein), mainly aimed at capturing *A. fraterculus* (Nava & Botton, 2010). For producers in integrated production systems, monitoring is mandatory and a control strategy is only carried out or recommended once the number of insects captured reaches a limit, known as the control level (Kovaleski & Ribeiro, 2003). The control level can vary according to the crop, the most commonly adopted being 0.5 flies/trap/day (Carvalho, 2005a).

In the south of Brazil, as in other regions, fruit fly control in orchards is usually carried out with insecticides applied as toxic baits or as a top dressing (Kovaleski *et al.*, 2000).

The ban on their use or the requirement to reduce the chemical residues allowed by importing countries, coupled with the decrease or drop in registrations of some active ingredients (ANVISA, 2006; Oliveira & Lucchese, 2013), leaves producers with few alternatives for the chemical control of fruit flies. Most of the insecticides available on the market have little or no depth effect, making it difficult to control the pest inside the fruit, leaving the use of toxic bait which, until now, has proved to be efficient when carried out correctly (Harter *et al.*, 2010).

There are few products registered in Brazil for the control of fruit flies, and for *A. fraterculus* there are five active ingredients from the organophosphate group (phosmet, malathion, dimethoate, fenitrotione and methidathion), one pyrethroid (deltamethrin) and one neonicotinoid (acetamiprid) (AGROFIT, 2014). On the other hand, according to the same source, for *C. capitata*, the number of products is slightly higher, with four organophosphates (chlorpyrifos, dimethoate, phosmet and malathion), three pyrethroids (cypermethrin, fenpropathrin and deltamethrin), one spinosyn (spinosad) and one diphenyl ether (etofenproxy).

In this context, Brazilian farmers are faced with the challenge of replacing conventional management, in which synthetic chemical products are used, often prophylactically, with an alternative system, rationalizing the tools used to control pests and diseases, respecting the need for the products and the bioecological relationships of the agroecosystem (Protas & Sanhueza, 2003).

Alternatives to chemical control are scarce, and in organic farming, for example, the means available for fruit fly control, recommended and approved by research, include bagging fruit, collecting fruit in covered pits, applying spinosyns and other substances such as kaolin (Lipp & Secchi, 2002; Nora & Hickel, 2006; Fofonka, 2007; Silva *et al*, 2007; Botton *et al.*, 2012). Natural biological control is another alternative; however, further studies are needed to optimize the performance of these agents, especially parasitoids, given that low parasitism rates have been recorded in species inventories (Salles, 1996; Gattelli, 2006; Pereira-Règo *et al.*, 2013).

## 2.2 Native parasitoids of fruit flies in Brazil

The species *Doryctobracon areolatus* (Szépligeti), *Utetes anastrephae* (Viereck) (Hymenoptera: Braconidae) and *Aganaspis pelleranoi* (Brèthes) (Hymenoptera: Figitidae) are the most frequent throughout Brazil and occur from Argentina to northern Mexico (López *et al.*, 1999).

*Doryctobracon areolatus* is often reported as a dominant species in surveys carried out in several Brazilian states, such as São Paulo (Mendes *et al.*, 2007; Marinho *et al.*, 2009), Bahia (Carvalho, 2005b), Minas Gerais (Alvarenga *et al.*, 2009), Rio Grande do Norte (Araujo & Zucchi, 2002b) and RS (Gattelli, 2006; Nunes *et al.*, 2012; Pereira-Rêgo *et al.*, 2013). According to Sivinski *et al.* (1998), this parasitoid prefers third instar fly larvae, but can attack larvae in earlier instars in case of competition with other species.

In Brazil, *D. areolatus* has already been recorded on guava in Rio Grande do Norte (Araujo & Zucchi, 2002b); on coffee in Rio de Janeiro (Souza *et al.*, 2005); on guava, juà, seriguela, papaya, cassava, umbu and cajà in Minas Gerais (Alvarenga *et al.*, 2005; 2009), 2005; 2009); in acerola, carambola, guava, jambo d'àgua, jambo vermelho, mango, pitanga, seriguela and umbu-cajà in Bahia (Matrangolo *et al.*, 1998; Carvalho, 2004; 2005b; Bomfim *et al.*, 2010); in pitanga, guava, peach, guabiroba, pitanga, araçà, mango, papaya and loquat, among others in São Paulo (Mendes *et al.*, 2007; Marinho *et al.*, 2009; Souza-Filho *et al.*, 2009).

*Utetes anastrephae* is a parasitoid of third instar larvae and has the smallest ovipositor among the braconids that attack fruit flies (Matrangolo *et al.*, 1998; Sivinski *et al.*, 2000). They are usually found on fruit collected from the ground, small in size and at a stage of development.

advanced ripening or already decomposing. This species can be found from Florida to Argentina (Sivinski *et al.*, 2000). According to Aluja *et al.* (2013), inside the host, the first instar *larvae of U. anastrephae* manage to win the competition with *D. areolatus*, because their mandibles are larger. The same result was reported by Paranhos *et al.* (2012) in relation to *D. iongicaudata*.

In Brazil, *U. anastrephae* has already been recorded in São Paulo, on pear, pitanga and loquat (Marinho *et al.*, 2009; Souza-Filho *et al.*, 2009); in Bahia, on umbu-cajà, pitanga, carambola, guava, jambo d'àgua, jambo vermelho and seriguela (Matrangolo *et al., 1998; Carvalho, 2004; 2005b; Bomfim et al.*, 2010), 1998; Carvalho, 2004; 2005b; Bomfim *et al.*, 2010); in Minas Gerais, in cassava and umbu (Alvarenga *et al.*, 2009); in Rio de Janeiro, in coffee (Souza *et al.*, 2005) and in Rio Grande do Norte, in cajarana (Araujo & Zucchi, 2002b).

*Aganaspis pelleranoi* is a parasitoid of third instar larvae, which usually seeks its hosts in the cracks of the fruit, in order to get close to the larvae, since it has a short ovipositor (Sivinski *et al.*, 2000). Guimarâes *et al.* (1999) carried out a survey in all regions of Brazil, in various states, collecting fruit and recording the emergence of *A. pelleranoi*. The authors found this species in Amazonas, Bahia, Mato Grosso do Sul, Goiás, São Paulo, Rio de Janeiro and the three southern states. They also recorded that, among the 29 fruits related to *A. pelleranoi*, there are native ones, such as guava, pitanga, guabiroba and araçà, and exotic ones, such as peach. The species has also been recorded in Rio de Janeiro on coffee (Souza *et al.*, 2005), in Minas Gerais on umbu (Alvarenga *et al.*, 2009), in Bahia on acerola, carambola, guava and pitanga (Matrangolo *et al.*, 1998; Carvalho, 2005b) and in São Paulo on guava (Souza-Filho *et al.*, 2009).

In addition to these species, others occur in Brazil in less abundance and frequency, such as *Doryctobracon brasiliensis* (Szépligeti) in São Paulo (Marinho *et al., 2009;* Souza-Filho *et al., 2009),* Minas Gerais (Alvarenga *et al.*, 2009) and Rio de Janeiro (Souza *et al.*, 2005). In the latter state, Souza *et al.*

(2005) also observed the braconids *Asobara anastrephae* (Muesebeck), *Opius bellus* Gahan and *Doryctobracon fluminensis* Lima and the figitids, *Dicerataspis flavipes* (Kieffer) and *Odontosema anastrephae* Borgmeier. These same braconids were found in São Paulo by Souza-Filho *et al.* (2009) and Marinho *et al.* (2009).

In RS, to date, eight species of native parasitoid hymenoptera have been recorded, distributed in four families, Diapriidae with only one representative, *Trichopria anastrephae* Lima, as well as Pteromalidae, with *Pachycrepoideus vindemmiae* (Rondani), Figitidae with two species, *Odontosema albinerve* Kieffer and *A. pelleranoi* and Braconidae with *O. bellus, D. areolatus, D. brasiliensis* and *U. anastrephae* (Salles, 1996; Guimarâes *et al.*, 1999; Gattelli, 2006; Cruz *et al.*, 2011; Nunes *et al.*, 2012).

The species of Braconidae and *A. pelleranoi* parasitize only fruit fly *larvae*, while *P. vindemmiae* and parasitoids of the genus *Odontosema* parasitize both larvae and pupae (Salles, 1996). Species of the genus *Trichopria*, on the other hand, are parasitoids only of pupae (Wharton & Yoder, 2014). According to López *et al.* (1999), *P vindemmiae* is a generalist of Diptera larvae and may be less effective in pest control, because as its hosts become rarer, it may attack beneficial flies.

## 2.3 Biological control of fruit flies

The use of Braconidae hymenopterans has been prioritized for fruit fly control, due to their host specialization (Aluja *et al.*, 1990), which can be an advantage in classical biological control programs (Clausen, 1940).

Among the Braconidae, the opiines are cenobiont endoparasitoids, which oviposit in eggs or larvae of the host and emerge from the puparium (Wharton, 1989). Around 100 species of braconids have been recorded worldwide on fruit-infesting tephritids, the main ones being included in the genera *Diachasmimorpha* Viereck, *Psyttalia* Walker, *Utetes* Foster and *Opius* Wesmael (Wharton & Yoder, 2014).

The countries that have biological control programs for fruit flies have done so in the classic way and have used the inundative method, in which parasitoids are produced on a large scale in biofactories and released over large areas, with *D. longicaudata* being the most widely used species (Carvalho, 2006).

The releases that took place in the 1940s in Hawaii correspond to the first reported case of success with *D. longicaudata* (Carvalho *et al.*, 2000). Funasaki *et al.* (1988) reported that the parasitism rate of fruit flies rose from 15% to 47% after the parasitoid was released in that country.

According to Bess *et al.* (1961), more than 120,000 individuals of *D. longicaudata* were released in Hawaii between 1948 and 1950, along with 25 other species of Braconidae, mostly from South Asia, Indonesia and Malaysia. According to the authors, in 1949, *D. longicaudata* was responsible for half of the 25% of parasitism recorded in larvae present in guavas, a parasitism rate that reached more than 80% in 1951.

In Hawaii, control was sought for the fruit flies, *C. capitata*, *B. dorsalis* and *Dacus curcubitae* Coquillet (Diptera: Tephritidae), so several species of parasitoids were released, with *D. longicaudata* being used to control *B. dorsalis* (Funasaki *et al.*, 1988). However, this species was the one that best adapted to the three hosts and remained in the environment after the releases ended in the 1950s.

During this same period, Mexico was invaded by the same three species of fruit fly found in Hawaii and an alternative control method tested was the release of *D. longicaudata*, which adapted to the most frequent native species of fly, *Anastrepha ludens* (Loew) and increased parasitism from 20% to 60% (Cancino & Montoya, 2006). In an extensive study of fruit collection and parasitoid identification, according to Aluja *et al.* (1990), it was found that native Mexican parasitoids were responsible for only 10% of parasitism in species of the genus *Anastrepha*.

Aiming to repeat the success of Mexico, in Florida (USA), releases of *D.*

*longicaudata* began in 1972 in order to control *Anastrepha suspensa* (Loew), which had been accidentally introduced into this state in 1965 (Valle, 2006). The capture of flies in McPhail-type traps decreased from the fifth week after the first release of the parasitoid (Sivinski *et al.*, 1996) and the parasitism rate in *A. suspensa* reached 40%, with 95% occurrence of *D. longicaudata* (Carvalho & Nascimento, 2002).

In the 1970s, the governments of Mexico, Guatemala and the United States of America joined forces to control fruit flies, creating the Moscamed Trinational Program, which included, among other measures, the construction of biofactories and the rearing and release of *D. iongicaudata* to control *C. capitata* (Cancino & Montoya, 2006). Releases peaked between 1987 and 1988 in the Mazapa Valley, where 200,000 adults were released every week (Guillen *et al.*, 1991).

In 1992, Mexico began a national campaign to control the Mediterranean fruit fly. This campaign was based on six assumptions, according to Montoya *et al.* (2007): (1) constant monitoring with flycatcher lures; (2) cultural and mechanical practices to prevent or hinder fly attacks; (3) the application of insecticides selective to natural enemies; (4) the use of the sterile male technique; (5) the establishment of fly-free zones and (6) the release of *D. longicaudata*. According to the authors, success also depended on the partnership between: (1) the government and federal and state agencies and (2) field operators, producers, researchers and students. Today, the country is recognized as a Mediterranean fruit fly-free zone (Hernández, 2014).

*Diachasmimorpha longicaudata* was introduced in Costa Rica, Nicaragua, Panama, Argentina, Guatemala, Peru, Trinidad and Tobago and Brazil to control flies of the genera *Anastrepha, Bactrocera, Ceratitis* and *Dacus* (Ovruski *et al.*, 2000). According to the same authors, outside the Americas, this parasitoid has been successfully used in classical biological control programs in Australia, Fiji, Greece, Italy and Spain to control flies of the genus

*Bactrocera*.

In Costa Rica, *D. longicaudata* was introduced in 1958, along with eight other parasitoid species, to control *C. capitata*. After field studies, it was found that only *D. longicaudata* and *Biosteres arisanus* (Sonan) (Hymenoptera: Braconidae) remained in the environment, but with parasitism rates below 3% on *C. capitata* (Wharton *et al*., 1981).

In Argentina, releases were occasional, totaling 100,000 individuals of *D. longicaudata* between 1977 and 1986, in the province of Tucumàn (Schliserman *et al*., 2003). After the end of the releases and until 1999, according to the authors, no insects had been observed, however, between 2000 and 2002, 14 individuals were recovered.

In Brazil, *D. longicaudata* was only introduced in 1994, with specimens brought from Gainesville, Florida, by Embrapa Mandioca e Fruticultura Tropical, with the support of Embrapa Meio Ambiente's "Costa Lima" Quarantine Laboratory (Carvalho & Nascimento, 2002). The aim, according to the authors, was to study the species in order to assess the possibility of implementing a classic biological fruit fly control program in Brazil.

In the Recôncavo Baiano, 42,963 adults of *D. longicaudata* were released between August 1995 and July 1996 (Carvalho, 2005c). Subsequently, according to the author, this species was recovered from pupae originating from fruit collected in the region only until 1998. The introduction of this exotic parasitoid into the region did not cause the deletion of any of the main native species, and *D. areolatus, U. anastrephae, Opius* spp., *A. anastrephae* and *A. pelleranoi* were found both before and after the releases (Matrangolo *et al*., 1998; Carvalho, 2005c).

In the north of Minas Gerais, 34,000 pairs of *D. longicaudata* were released in a guava orchard between February and August 2003 (Alvarenga *et al*., 2005). According to the authors, the native parasitoid guild was not affected and the exotic remained in the environment, as 34 specimens were recovered from fruit collected in this orchard during the same period as the releases.

### 2.3.1 *Diachasmimorpha longicaudata* - bioecological aspects

The parasitoid *D. longicaudata* originated in the Indo-Australian region and tephritids of the genus *Bactrocera* are its original hosts (Carvalho & Nascimento, 2002).

The adult has a light brown to reddish brown body; the eyes are brown; the antennae are filiform and longer than the body and the wings are hyaline; the females measure 3.6 to 5.4 mm in length, not including the ovipositor which is longer than the whole body; the males measure 2.8 to 4.0 mm (Thompson, 1989). Chemoreceptors located in the ovipositor of *D.*

*longicaudata* allows the conditions of the oviposition substrate to be assessed, as well as the substances present in already parasitized larvae, in order to avoid over- and multiparasitism (Greany *et al.*, 1977). As well as assessing parasitism, this insect can also identify whether its possible host is a tephritid, because although it is a generalist parasitoid, it does not parasitize individuals from other Diptera families (Sthul *et al.*, 2011).

Firstly, the location of the fruit, which may contain their hosts at the ideal stage of development for oviposition, is achieved through the perception of products released by fermentation (Thompson, 1989). Once in the fruit, in simplified terms, the larvae are located by the vibrations they emit when feeding on the fruit pulp (Carvalho, 2005b).

The process of locating oviposition sites, however, is complex and can also be influenced by the size, color and species of the fruit that houses the host larva (Leyva *et al.*, 1991; Jang *et al.*, 2000; Segura *et al.*, 2012). According to Leyva *et al.* (1991), the same fruit that proves to be more attractive in olfactometry tests, in which smell is the only variable allowed, can be rejected when there is the possibility of choosing between visual and tactile stimuli.

Eben *et al.* (2000) observed in an olfactometer that infested fruit were more attractive to *D. longicaudata* females than uninfested fruit. Apparently, the injury caused to the fruit by the larvae may be responsible for the release of volatiles that are more perceptible than the larvae themselves (Segura *et al.*, 2012).

Studying the species *Diachasmimorpha juglandis* (Muesebeck), Henneman *et al.* (2002) found that females seek out damaged fruit more than healthy fruit when looking for oviposition sites.

Because of the tritrophic interaction that takes place in a case of fruit fly parasitism, olfactory stimuli relating to the species of fruit, infestation, injury and decomposition, as well as visual clues and vibrations emitted by the larva, can influence the decision to choose (Jang *et al.*, 2000). According to the authors, the parasitoid has to process various pieces of information about the options for oviposition sites in order to guarantee the maximum chance of success.

*Diachasmimorpha longicaudata* is a species with facultative parthenogenetic reproduction (Carvalho, 2003). The larva of the parasitoid hatches before the pupal stage of the host, obligatorily killing it inside the puparium, where it carries out the instar changes (Lawrence, 1981).

The biological cycle of *D. longicaudata* at 25 °C is approximately 19 days for females and 18 for males, regardless of whether the host is *C. capitata* or *A. fraterculus* (Meirelles *et al.*, 2013). According to the authors, the average longevity of *D. longicaudata*, when reared on these two hosts, is around 25 days.

According to Greany *et al.* (1977), *D. longicaudata* can copulate from the first day of life and, from the second day, the females are able to lay fertile eggs. However, according to Meirelles *et al.* (2013), in the first two days, most of the offspring are males. In the laboratory (25 ± 2°C; 65 ± 10% RH; 14 hours of photophase), these authors recorded similar average daily fecundity in *A. fraterculus* larvae (6 eggs/female/day) and in *C. capitata* (5 eggs/female/day), as well as average fertility in *A. fraterculus* (65%) and *C. capitata* (72%).

In laboratory tests, Meirelles (2011) found that the basal temperature (Tb) for *D. longicaudata* reared on *A. fraterculus* was 12.5 °C, while on *C. capitata* it was 7.8 °C. According to the author, the optimum temperature range for the development of the parasitoid reared on the South American fruit fly is between 16.3 and 22.4 °C and between 11.5 and 21.8 °C for the Mediterranean fruit fly. It is important to note that these are the two most important tephritid species in RS (Kovaleski *et al.,* 2000) and that temperatures below the optimum range and even below Tb are common in winter, early spring and late fall (between May and October) in southern Brazil (INMET, 2014).

# 3. MATERIAL AND METHODS

## 3.1 Insect breeding

The insects used in the experiments are kept in the Biology, Ecology and Biological Control of Insects Laboratory (BIOECOLAB) of the Agronomy Faculty of UFRGS (FAGRO), in an air-conditioned room (25 ± 2 °C; 65 ± 10% RH; 14 hours of photophase) or in an air-conditioned chamber (25 ± 2 °C; 65 ± 10% RH; no photophase).

### 3.1.1 *Diachasmimorpha longicaudata*

Breeding has been carried out according to the methodology proposed by Carvalho *et al.* (1998). The adult cages are made of wood, covered in voile fabric (60 cm x 25 cm x 50 cm) and can hold up to 400 individuals. The water is supplied by capillarity through a strip of Spontex Resist® fabric, the base of which is immersed in the water, arranged in a plastic pot with a perforated lid. The gelatinous food consists of water (120 mL), honey (120 mL), agar (0.8 g), ascorbic acid (0.05 g) and nipagin (0.005 g). The agar, dissolved in hot water, is mixed with the other ingredients. The diet, while still warm, is poured (2 mL) into Petri dishes (0 5 cm), which are kept at room temperature until the liquid hardens. They are then protected with PVC film and kept in a refrigerator (± 4 °C) for up to 10 days. Each cage receives a plate of food, which is changed twice a week.

Every day, the parasitoid cages receive an oviposition unit (Figure 1) with third instar *A. fraterculus* larvae, approximately nine days old, for half an hour. The number of larvae is approximately 400, trying to maintain a ratio of two larvae per female.

FIGURA 1. Oviposition unit, made of acrylic plates, for offering *Anastrepha fraterculus* larvae to *Diachasmimorpha longicaudata* females.

Each cage can receive up to three oviposition units per day. The units were made using a Gerbox lid (11 cm x 11 cm) with a circular opening (0 9 cm) protected with voile fabric glued to the outside of the lid. The larvae are placed inside the opening, on the voile fabric and the lid is pressed against the bottom of a Gerbox, preventing the larvae from getting out, leaving them exposed for parasitism.

After exposure, the larvae are placed in a Gerbox with sand at the bottom and kept in an air-conditioned chamber (25 ± 1°C; 60 ± 10% RH; no photophase) until the parasitoids emerge.

### 3.1.2 *Anastrepha fraterculus*

The adults are kept in wooden cages (45 cm x 30 cm x 30 cm), covered with voile fabric. The adults' food consists of crystal sugar, brewer's yeast and Bionis® hydrolyzed yeast (in a 3:1:1 ratio). In addition to these ingredients, one vitamin complex tablet (DiariumR - VitaminLife) is macerated and mixed into every 250 g of diet. The food is offered in Petri dishes (0 5 cm) and changed once a week.

The flies' oviposition substrate is in the shape of a 30 cm x 30 cm bag, made from blue voile fabric covered with silicone and with an opening at one end with a lid, through which water is poured (Figure 2). This substrate is placed on top of the cage and the flies oviposit through the fabric. The eggs are immersed in water, which is replaced daily. The eggs removed are placed on filter paper in glass Petri dishes (0 9.5 cm). These plates are wrapped in newspaper to keep the eggs in the dark and kept in an air-conditioned chamber (25 ± 1°C; 60 ± 10% RH; no photophase) for two days.

The larvae's diet, proposed by Téran (1977), consists of raw peeled carrots (125 g), cooked peeled carrots (125 g), crystal sugar (125 g), brewer's yeast (25 g), water (175 mL), nipagin (1.1 g), citric acid (3.6 g) and sodium benzoate (1.1 g). The ingredients are mixed in a blender and the mixture is distributed in square Styrofoam trays (15 cm x 15 cm x 2 cm) on which the filter papers with the eggs are placed. These are covered with another tray, wrapped in newspaper and placed in an air-conditioned chamber (25 ± 1°C; 60 ± 10% RH; no photophase) for seven days. After this period, the trays are taken back to the climate chamber and placed in plastic containers (50 cm x 30 cm x 5 cm), on a layer of sterilized sand and covered with voile fabric. After nine days, the containers are removed from the chamber, the sand is sieved and the pupae are separated. These are transferred to plastic jars (250 mL) with a screened lid, containing sterilized sand at the bottom and kept in the same chamber for around 10 days. A new cage is set up with approximately 400 pupae and can be kept for up to 50 days.

FIGURA 2. Oviposition substrate for *Anastrepha fraterculus* on the rearing cage.

**2.2 Evaluation of infestation and parasitism in the field and the increase**

**in parasitism after exposure to *Diachasmimorpha longicaudata* in the laboratory**

Whole fruits were collected, on the ground, under the canopy of fruit trees in three locations: UFRGS Agricultural Experimental Station (EEA-UFRGS) (30°06'16"S and 51°39'58"O), in Eldorado do Sul: loquats (*Eriobotrya japonica* Lind.) and peaches (*Prunus persica* L.) (Rosaceae), persimmons (*Diospyros kaki* L. F.) (Ebenaceae) and araçâs (*Psidium cattleianum* Sabine) (Myrtaceae); Centro Agricola Demonstrativo (CAD) (30°07'14"S and 51°05, 1ΓO), in Porto Alegre: guabirobas (*Campomanesia xanthocarpa* Berg) (Myrtaceae) and at the UFRGS Agronomy Faculty (FAGRO) (30°06'98"S and 51°14'19"O): guavas (*Psidium guajava* L.), mountain guavas (*Acca sellowiana* O. Berg) and pitangas (*Eugenia uniflora* L.) (Myrtaceae).

The fruit was taken to the laboratory, weighed, counted and washed with 1% hypochlorite. After drying, they were divided into two equal batches, one exposed to *D. longicaudata* and the other not (control). For exposure, the fruits were kept on the floor of wooden cages (50 cm x 50 cm x 60 cm) covered with voile fabric, where couples of *D. longicaudata*, aged between five and ten days, were released in the proportion of one couple for each group of ten fruits, for 24 hours. After this, the fruit was placed individually in 140 mL plastic jars for guabiroba, araçà, pitanga, loquat and guava and 2 L jars for guava, persimmon and peach, containing vermiculite at the bottom, and kept in an air-conditioned chamber (25 ± 2 °C; 65 ± 10 % RH; 14 hours of photophase). The same procedure was carried out with the control batch, but without contact with the parasitoid. After seven days, the vermiculite was sieved and the pupae separated. This procedure was repeated daily until no pupae had been recovered.

The pupae obtained from both batches were individualized in identified plastic jars (140 mL) and kept in an air-conditioned chamber (25 ± 1°C; 60 ± 10% RH; no photophase) until the flies or parasitoids emerged. The insects were killed and placed in eppendorfs (1.5 mL) with 70% alcohol for later identification. All pupae from which no emergence was recorded were dissected after 20 days, looking for the presence of flies or parasitoids. These were identified using dichotomous keys (Zucchi, 2000) and the BIOECOLAB reference collection. The parasitoid identifications were confirmed by Dr. Miguel Francisco de Souza Filho and Dr. Valmir Antônio Costa, from the Biological Institute of Campinas, SP.

Infestation intensities, parasitism rates, numbers and species of both flies and parasitoids were recorded before and after exposure to *D. longicaudata.*

The parasitism indices were calculated for each repetition. The apparent parasitism index, or parasitism success, took into account only the emerged insects and was calculated as follows: apparent parasitism = [n° parasitoids/ (n° parasitoids + n° flies)] x 100.

The average number of flies and pupae were subjected to the Shapiro-Wilk homoscedasticity test. Differences in the average number of fruit flies and pupae between treatments were compared using the Mann-Withney test. The BioEstat 5.0® program (Ayres *et al.*, 2007) was used for statistical analysis.

### 2.3 Parasitism *of Diachasmimorpha longicaudata* in semi-field conditions

This experiment used seedlings of native myrtle trees, guabirobeira, guava and araçazeiro, and exotic species, nespereira, peach and kunquateiro [*Fortunella margarita* (Lour.) (Rutaceae)]. The seedlings were between two and four years old and were all purchased from commercial nurseries six months before the experiment began. Until the start of the work, the seedlings remained in the experimental area of the Plant Health Department, at the Agronomy Faculty of UFRGS, under an 8-m$^2$ roof, protected on the sides and top with shade (retaining 70% of the light), which allowed rain to pass through. The plants were irrigated daily when it wasn't raining and all of them received foliar fertilization (Biofert Universal®) at the concentration recommended by the manufacturer (5 mL/L) every 15 days using a hand sprayer, just after irrigation. Fertilization was suspended one week before the start of the experiment.

The fruit used in the experiment, from each of the seedling species, had the fruit fly infestations controlled, which took place from June 2012 to February 2013, according to the fruiting period of each one. The guabirobas were harvested at the Porto Alegre Demonstrative Agricultural Center (CAD), the araçâs*, kumquats and* guavas at the UFRGS Agronomy Faculty Campus and the peaches and loquats at the UFRGS Agronomy Experimental Station.

At the beginning of fruiting, the fruit was protected on the trees with cages made from voile, each one trying to house more than one fruit, preventing oviposition by flies in the field. The voile cages were rectangular in shape, 30 cm long by 20 cm wide, with an opening at one end that could be closed with a rope. The cages were attached to the stem of the plant, with the fruit(s) inside.

At the beginning of ripening for each of the species, *A. fraterculus* females aged between 12 and 20 days were introduced into each cage, in the proportion of two individuals per fruit. The females remained in contact with the fruit for 48 hours, when they were removed. The fruit remained protected and was harvested nine days after the removal of the females, when the larvae should have been in the third instar.

The fruit collected from the plants was placed in two wooden cages (control and treatment) (2 m x 2 m x 2 m), covered with voile fabric, kept in a greenhouse (6 m x 4 m) with a transparent plastic cover and shingled sides, located in the experimental area of FAGRO's Plant Health Department. Four seedlings of each of the species described above were placed in each cage, which were approximately 120 cm high. The plants were randomly distributed in the cage (Figure 3).

In the cages, the harvested fruit was placed in a Petri dish lid (0.9 cm) containing vermiculite, which was placed on top of the pot, next to the base of the seedling of the same species as the fruit being evaluated. In one of the cages containing the seedlings, these fruits were exposed to four non-experienced couples of *D. iongicaudata*, five to seven days old, from the BIOECOLAB breeding facility, in which males and females were kept together since emergence. The parasitoids remained in the cage with the fruit for 24 hours and were then discarded. Fruits of each species were exposed on three consecutive days, using eight fruits/day. At each exposure, the parasitoid couples were replaced and the seedlings were rotated in the cage.

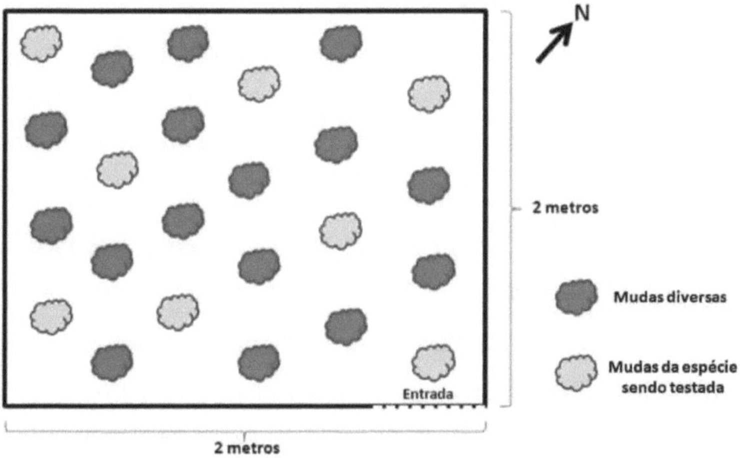


FIGURA 3. Sketch showing the distribution of fruit seedlings inside the cage used in the experiment under semi-field conditions.

In the other cage containing the fruit seedlings (control), only the fruit infested with fruit flies was placed, following the same protocol described above.

As not all the fruits were used on the same day, the infested ones that were going to be exposed on the following days were kept in plastic jars with lids and kept in the same greenhouse as the cages until they were exposed.

The fruits removed from each cage after the exposure period were stored individually in plastic jars with vermiculite on the bottom and a lid with an opening protected with voile and kept in an air-conditioned chamber (25 ± 2 °C; 65 ± 10% RH; 14 hours of photophase). From the seventh day onwards, the vermiculite was sieved daily to collect the pupae and the fruit was placed back in the jars. This operation was repeated until the tenth day, when nothing was recovered from the samples. The pupae were individualized and kept in the same chamber until the flies and/or parasitoids emerged. The average fruit fly infestation, total and average number of pupae per fruit and parasitism indices for each fruit species were recorded.

The apparent parasitism index was calculated for the group of exposed fruit. The calculation of the parasitism indices, the homoscedasticity test and the software for statistical calculation have already been described in section 3.2. The average number of flies per fruit and pupae per fruit was compared between treatments using the t-test ($\alpha = 0.05$), with the exception of those from guavas, where the Mann-Whitney test was used.

### 3.4 Evaluation of fruit fly parasitism in the field following releases of *Diachasmimorpha longicaudata*

The field experiments were carried out at the UFRGS Agronomic Experimental Station (30°06'16"S and 51°39'58"O). The releases of *D. iongicaudata* were made in loquat (0.15 ha), peach (1.17 ha) and coconut (0.74 ha) orchards, which are collections containing more than one variety and/or cultivar of each fruit, and in an area of 0.3 ha of araçá trees which were not planted with the intention of forming an orchard, but grew spontaneously (Figure 4).

The area of araçazeiro trees was 20 m away from the caquizeiro orchard. Both were to the north of the study area. The loquat orchard was 100 m south of the coconut trees. The peach trees were 20 m to the east of the loquat trees. There was approximately 400 m between the southernmost end of the peach orchard and the northern end of the araçazeiro area (Figure 4).

In the years in which the study was carried out, management was different in each orchard. No herbicides were used. The areas of loquat, peach and persimmon trees were mowed at pruning, thinning and harvesting times, as necessary. The araçazeiro trees did not receive any phytosanitary treatment, nor did they receive any pruning, thinning or mowing.

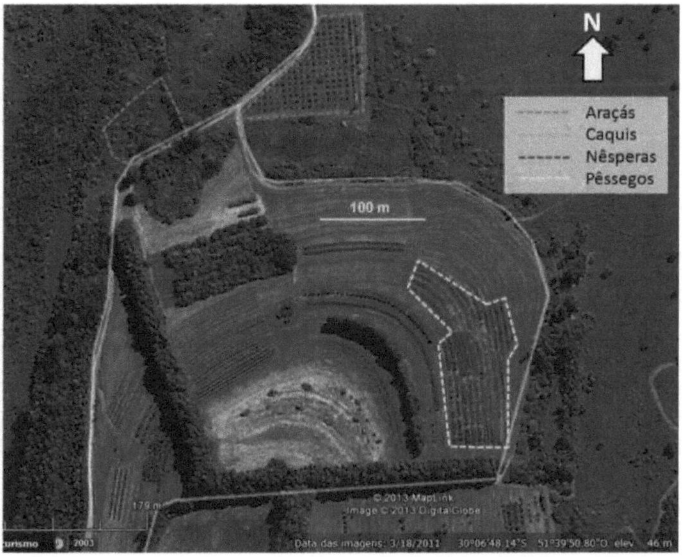


FIGURA 4. Satellite image of the experimental area, Eldorado do Sul, RS (2011/2012), delimiting in color the areas where each fruit species was collected and where *Diachasmimorpha longicaudata* was released (Google Earth, 2011).

In the peach orchard, established since 1999, *Grapholita molesta* (Busck) (Lepidoptera: Tortricidae) is controlled with phosmet (Imidan 500 WP®), once a season, and applications of the pheromone-based product Splat Grafo®. To control fruit flies, applications are made of a toxic bait [50 g of dimethoate (Nortox®500 EC) + 100 L of water + 5 kg of sugar], which is sprayed weekly on plants at the edge of the orchard. Applications begin in the second half of September, the final half of the loquat harvest.

The loquat orchard, planted in 2003, only receives weekly applications of the same toxic bait for fruit flies, starting in September, when the last fruit is being harvested. The coconut plantation, set up in 1994, doesn't receive any insecticide applications, only sulphocalcic syrup at the start of the crop's vegetative phase.

In the first year of the experiment (2011/2012 harvest), considered to be pre-release, the ripe fruits of all the species were collected on the ground under the canopy of the plants, with the exception of the loquats, which were collected ripe on the plant.

The fruits were taken to the laboratory, counted, weighed, washed and placed in plastic trays with vermiculite on the bottom, covered with fleece, where they remained at room temperature. Seven days later, the vermiculite was sieved to separate the pupae, which were kept in plastic jars (140 mL) until they emerged. The fruit was returned to the trays and the procedure was repeated weekly until no pupae were collected. The insects that emerged were killed and identified using dichotomous keys (Zucchi, 2000) and the BIOECOLAB reference collection. Confirmations of the parasitoid identifications were made by the same experts mentioned in section 3.2.

In the second year (2012/2013 harvest), approximately 1,700 parasitoids/ha, in the ratio of one male to one female, were released every fifteen days on average. The releases began when the apples reached the semi-ripe stage (in the first half of August) and continued until the end of the coconut crop (in the second half of March). A total of 10 releases were made, only when the orchards had ripe fruit. No parasitoids were released in the off-season.

The releases were always carried out in the morning, near the tops of the plants that had fruit. The cages containing the adults were held in the hand, with the door open, and the area was walked through until all the parasitoids had left. In the second year, in addition to the releases, fruit was collected, repeating the methodology used in the pre-release year.

In the third year (2013/2014 harvest), no parasitoids were released, but fruit was collected from the four fruit trees following the same procedure as in the first year.

The species of flies and parasitoids present on the fruit in the three years evaluated were recorded.

Parasitism and infestation rates were calculated for all three years and for all fruit trees. Infestation was calculated by dividing the number of pupae obtained by the number of fruits sampled. The infestation data (number of pupae/kg of fruit) and parasitism indices before and after the releases (increase in parasitism) were compared using the chi-square test of heterogeneity ($\alpha$ = 0.05). The same software referred to in items 3.2 and 3.3 was used for statistical analysis.

# 4. RESULTS AND DISCUSSION

## 4.1 Evaluation of infestation and parasitism in fruit brought in from the field and the increase in parasitism after exposure to *Diachasmimorpha longicaudata* in the laboratory

### 4.1.1 Fruit fly infestation and species

*Anastrepha fraterculus* was the only tephritid recorded on araçâs, guabirobas, pitangas, peaches and guavas and the most abundant on loquats and guavas in both the control fruit lots and those exposed to *D. longicaudata*. The prevalence of the South American fruit fly was expected given that it was also the only species found on pitangas and guavas in the Vale do Cai region, RS, by Meirelles *et al.* (2005) and Meirelles *et al.* (2006). In the same region, Gatelli *et al.* (2008) observed that, apart from pitanga and guava, this was the only Tephritidae present in guabirobas, araçâs and guavas. In Pelotas, Cruz *et al.* (2011) found only this species of fly on guavas; however, it is worth mentioning that the authors collected 54 fruits, while in the present study this number was four times higher. Individuals of *A. fraterculus* were also the only ones observed by Pereira-Rêgo *et al.* (2013) on araçâs, guavas and guava-serranas from Porto Alegre and Bento Gonçalves, RS.

One specimen of *C. capitata* was found in the test plot of loquats and one in guavas. In persimmons, on the other hand, the Mediterranean fruit fly was the only species recorded. Similarly to this study, Nunes *et al.* (2012) found that the South American fruit fly was the most abundant on araçâs, pitangas, guavas, serranas guavas, loquats and peaches, and the Mediterranean fruit fly was the only one observed on persimmons in Pelotas and Capào do Leâo, RS.

Lonchaeidae flies were recorded on guavas, serrano guavas and loquats, both in the exposed fruit group and in the control group. In the araçâs, individuals from this family were only present in the control group, in which only two adults were recovered. In the Cai Valley, lonqueids had already been found on araçâs, guavas and guava-serranas by Gatelli *et al.* (2008) and, in São Paulo, on persimmons, loquats and peaches by Raga *et al.* (2011). In addition to the fruit trees mentioned by the previous authors, Nunes *et al.* (2012) also found Lonchaeidae on pitangas.

However, among the fruit trees sampled, only in caquize trees was *C. capitata* the most abundant fly species (and the only one recorded). It can be concluded that *A. fraterculus* is also the most abundant and frequent tephritid in the studied regions of Porto Alegre and Eldorado do Sul.

The pattern of occurrence of the fruit fly species recorded in this study, which occurs in a large part of Rio Grande do Sul, seems to be reversed in the western region of the state, close to the borders with Argentina and Uruguay. This is pointed out in the work of Dias *et al.* (2013) who found *A. fraterculus* on loquats, guavas and pitangas, however, the number of *C. capitata* found on these fruits was 800% higher on guavas and 300% on loquats, this being the only species observed on persimmons and peaches.

Considering the intensity of fly infestation on the fruit, the average number of pupae per fruit was statistically lower in the group exposed to the exotic parasitoid on loquats (U = 2989.5; P < 0.0001), pitangas (U = 16846.5; P = 0.0064) and araçâs (U = 39499; P = 0.0096) and did not differ in peaches (U = 1167.5; P = 0.569), guabirobas (U = 19071.5; P = 0.422), guavas (U = 4305; P = 0.089) and persimmons (U = 1096.5; P = 0.29) (Table 1).

TABLE 1. Total number of fruit collected, infestation, average number of flies per fruit and parasitism index in groups of fruit with and without (control) exposure to *Diachasmimorpha longicaudata*, collected in Eldorado do Sul and Porto Alegre, RS. (August/2011 to April/2012).

| **Eldorado do Sul** | | | | |
|---|---|---|---|---|
| | **Loquats** | | **Peaches** | |
| | Witness | Exposed | Witness | Exposed |
| **FRUIT** | 100 | 100 | 50 | 50 |
| **Pupae/fruit** | 4,7 ± 0,39 A* | 2,4 ± 0,26 B | 0,6 ± 0,10 A | 0,4 ± 0,07 A |
| **MOSCAS/FRUTO** | 3,4 ± 0,30 A | 1,0 ±0,158 B | 0,44 ± 0,092 A | 0,14 ± 0,049 B |
| **PARASITISM INDEX (%)** | 0,28 | 14,75 | 0 | 30 |
| | **Khakis** | | **Araçâs** | |
| | Witness | Exposed | Witness | Exposed |

| | | | | |
|---|---|---|---|---|
| **FRUIT** | 50 | 50 | 300 | 300 |
| **Pupae/fruit** | 0,3 ± 0,07 A | 0,1 ± 0,05 A | 2,4 ± 0,10 A | 2,1 ± 0,11 B |
| **MOSCAS/FRUTO** | 0,16 ± 0,053 A | 0,06 ± 0,034 A | 1,4 ± 0,08 A | 0,44 ± 0,047 B |
| **PARASITISM INDEX (%)** | 0 | 0 | 6,25 | 68,4 |
| **Porto Alegre** | | | | |
| | **Guabirobas** | | **Mountain guavas** | |
| | Witness | Exposed | Witness | Exposed |
| **FRUIT** | 200 | 200 | 100 | 100 |
| **Pupae/fruit** | 0,9 ± 0,08 A | 0,7 ± 0,007 A | 2,6 ± 0,18 A | 2,2 ± 0,21 A |
| **MOSCAS/FRUTO** | 0,49 ± 0,055 A | 0,27 ± 0,039 B | 1,72 ± 0,138 A | 0,71 ± 0,099 B |
| **PARASITISM INDEX (%)** | 8,4 | 36,8 | 7,4 | 59,5 |
| | **Pitangas** | | **Guavas** | |
| | Witness | Exposed | Witness | Exposed |
| **FRUIT** | 200 | 200 | 100 | 100 |
| **Pupae/fruit** | 0,4 ± 0,03 A | 0,2 ± 0,03 B | 2,6 ± 0,12 A | 2,3 ± 015 A |
| **MOSCAS/FRUTO** | 0,32 ± 0,034 A | 0,27 ± 0,019 B | 1,8 ± 0,11 A | 0,86 ± 0,084 B |
| **PARASITISM INDEX (%)** | 0 | 25,5 | 9 | 59,5 |

* Values followed by different letters between groups of control fruit and fruit exposed to the parasitoid differ significantly by the Mann-Whitney test ($\alpha = 0.05$).

The reduction in the average number of pupae found in the first three species of fruit may indicate that some flies died in the larval or pupal stages as a result of the oviposition activity of *D. longicaudata*. This fact was also pointed out by Ashley *et al.* (1976) who reported that batches of larvae exposed to *D. longicaudata* showed up to 33% higher mortality. In work with the braconids *Fopius arisanus* (Sonan) and *Psyttalia fletcheri* (Silvestri), Harris *et al.* (2010) warned that parasitism may be underestimated if the mortality of the hosts in the young stage is not taken into account. The mortality of the fly, without the parasitoid completing its life cycle, reduces the chances of the biological control agent establishing itself in the environment. On the other hand, flies die without reaching the adult stage and causing damage, and the fact that they leave no offspring will contribute to their population reduction

### 4.1.2 Parasitism and parasitoid species

No parasitoid specimens were recovered from both groups of fruit (control and exposed). In peaches and pitangas, no parasitoids were recorded in the control group. The parasitism rates and species found varied in the other fruit groups (Table 1).

The absence of parasitoids in persimmons, as well as in the control group of peaches, may be related to the low fruit fly infestation observed in these two fruit crops. However, this is not the only factor, as guabirobas also had a low infestation, but the parasitism index was the second highest recorded in the study (Table 1).

Because they are exotic species, the cacao and peach trees are not evolutionarily associated with *A. fraterculus* or its parasitoids, which would partly explain the results obtained. However, in peach trees in RS, the presence of native parasitoid species, such as *D. areolatus* and *P vindemmiae* (Salles, 1996) and *D. areolatus* and *A. pelleranoi* (Nunes *et al.*, 2012) was found, at rates of between 1.1% and 1.26%, which is therefore low. In this same fruit tree, in São Paulo, Guimaraes *et al.* (1999) recorded the figitoids *A. pelleranoi* (1.58%), *L. anastrephae* (2.17%), *O. anastrephae* (0.42%) and *Trybliographa* sp. (0.08%), at similarly low rates. Thus, the hypothesis that native parasitoids do not recognize hosts in peaches does not fully support the results of this study. A greater sampling effort might have made it possible to visualize fruit fly parasitism in this fruit species, since 100 peaches from each treatment were examined.

The absence of parasitoids on the peaches may also be associated with the use of insecticides to control fruit flies, through the application of toxic baits in the orchard where the study was carried out. Although the toxic bait was only sprayed on fruitless plants (Item 3.3), the parasitoids may have visited the treated areas in search of food. In addition, there is the possibility of drift, whereby ripe fruit would end up receiving the insecticide. In addition, applying baits on a weekly basis may reduce the fruit fly population in the orchard, thus reducing the supply of hosts for the parasitoids.

Another aspect that can be pointed out is the thickness of the flesh of these fruits. According to Hickel (2002), in fruits with thick flesh (> 3.39 mm), parasitism rates by Braconidae individuals tend to be lower, due to the difficulty of finding their hosts. However, among the thick-fleshed fruits collected, guava and guava-serrana showed high infestation and parasitism rates compared to the others, except that these are species native to Latin America and well adapted to the South American fruit fly and its parasitoids.

In the case of the pitangas, the absence of parasitism in the control group can be attributed to the incidence of rust (*Puccinia psidii*), which left the fruit mummified or with a corky appearance, with hard skin and flesh (Glen *et al.*, 2007). This disease may have left pitangas, which are small fruits, with little or no surface free of "wounds", making it difficult for both fruit flies and parasitoids to act.

On the other hand, in the batch of pitangas exposed to the parasitoid, the wounds do not seem to have prevented the action of *D. longicaudata*. The high pressure of parasitism to which these fruits were subjected, one female for every 10 pitangas, for 24 hours, in a 150,000 cm cage[3] and the absence of other hosts for oviposition, meant that the parasitism rate went from 0% to 25.5% (Table 1). In the field, this would probably not have happened, as the female parasitoids would have been free to look for fruit in better conditions to oviposit, which seems to have happened if we look at the results obtained in the control group.

The parasitism rate found in guabiroba in the control batch in this study was the second highest among the fruit collected in Porto Alegre (Table 1), although it was still lower than that found by Salles (1996) (9.5%) and Gatelli (2006) (25.8%).

Among the native parasitoid species, *D. areolatus* was present in all the fruit trees whose control plots showed parasitism (Table 2), *U. anastrephae* was only found in loquats and *A. pelleranoi,* although it did not occur in the control plots of guabirobas and guava-serranas, was recorded in the treatment group of these two fruit trees.

TABLE 2 - Total number of parasitoids (N) and relative frequency (%) (fr) recorded in the group of fruit not exposed to *Diachasmimorpha longicaudata* (control) and in the exposed group, in fruit collected in Eldorado do Sul and

Porto Alegre, RS.

| Parasitoids | Eldorado do Sul | | | | | | | |
|---|---|---|---|---|---|---|---|---|
| | Loquats | | | | Peaches | | | |
| | Witness | | Exposed | | Witness | | Exposed | |
| | N | fr | N | fr | N | % | N | fr |
| **Natives** | 1 | 100 | 0 | | 0 | | 0 | |
| *Doryctobracon areolatus* | 1 | 100 | | | | | | |
| *Utetes anastrephae* | | | | | | | | |
| *Aganaspis pelleranoi* | | | | | | | | |
| **Exotic** | | | | | | | | |
| *Diachasmimorpha longicaudata* | 0 | | 18 | 100 | 0 | | 3 | 100 |

| Parasitoids | Araçâs | | | | Khaki | | | |
|---|---|---|---|---|---|---|---|---|
| | Witness | | Exposed | | Witness | | Exposed | |
| | N | fr | N | fr | N | % | N | fr |
| **Natives** | 28 | 100 | 22 | 6,1 | 0 | | 0 | |
| *Doryctobracon areolatus* | 10 | 35,7 | 9 | 2,5 | | | | |
| *Utetes anastrephae* | 6 | 21,4 | 1 | 0,3 | | | | |
| *Aganaspis pelleranoi* | 12 | 42,9 | 12 | 3,3 | | | | |
| **Exotic** | | | | | | | | |
| *Diachasmimorpha longicaudata* | 0 | | 338 | 93,9 | 0 | | 0 | |

| Parasitoids | Porto Alegre | | | | | | | |
|---|---|---|---|---|---|---|---|---|
| | Guabirobas | | | | Mountain guavas | | | |
| | Witness | | Exposed | | Witness | | Exposed | |
| | N | fr | N | fr | N | fr | N | fr |
| **Natives** | 9 | 100 | 12 | 23,5 | 14 | 100 | 15 | 23,4 |
| *Doryctobracon areolatus* | 1 | 11,1 | | | 3 | 21,4 | 4 | 6,3 |
| *Utetes anastrephae* | 8 | 88,9 | 10 | 19,6 | 11 | 78,6 | 8 | 12,5 |
| *Aganaspis pelleranoi* | | | 2 | 3,9 | | | 3 | 4,7 |
| **Exotic** | | | | | | | | |
| *Diachasmimorpha longicaudata* | 0 | | 39 | 76,5 | 0 | | 49 | 76,6 |

| Parasitoids | Pitangas | | | | Guavas | | | |
|---|---|---|---|---|---|---|---|---|
| | Witness | | Exposed | | Witness | | Exposed | |
| | N | fr | N | fr | N | fr | N | Fr |
| **Natives** | 0 | | 0 | | 18 | 100 | 19 | 14,8 |
| *Doryctobracon areolatus* | | | | | 6 | 33,3 | 5 | 3,9 |
| *Utetes anastrephae* | | | | | 5 | 27,8 | 4 | 3,1 |
| *Aganaspis pelleranoi* | | | | | 7 | 38,9 | 10 | 7,8 |
| **Exotic** | | | | | | | | |
| *Diachasmimorpha longicaudata* | 0 | | 24 | 100 | 0 | | 109 | 85,2 |

Medlar was the only exotic fruit to show parasitism by a native species, represented by just one individual of *D. areolatus*. In this same fruit, this parasitoid had already been found in the Pelotas region by Salles (1996) and Nunes *et al.* (2012), in São Paulo by Marinho *et al.* (2009) and in Argentina by Schliserman *et al.* (2010). Compared to the rates recorded by Salles (1996) (7.8%), Marinho *et al.* (2009) (< 5%), Schliserman *et al.* (2010) (< 5%) and Nunes *et al.* (2012) (0 to 3.08%), the rate found in this study in the control plot was low (Table 1). However, as it is a plant that bears fruit at the end of winter, it is a key host for fruit fly management and also for the multiplication of parasitoids, which could act to prevent fruit fly populations from reaching high levels in late spring and summer, when the main fruit trees in RS ripen and are harvested.

In guabiroba, *D. areolatus* and *U. anastrephae* were collected in the control plot, the latter being the most frequent (88.9%) (Table 2). In the group exposed to *D. longicaudata, D. areolatus* was not recorded and *A. pelleranoi* was only observed in the treatment plot.

A situation similar to that observed in guabirobas was found in the control fruit group, with the presence of only *D. areolatus* and *U. anastrephae*, the latter also being the most frequent (78.5%). This is the first record of *U. anastrephae* on guava in RS. The closest region to the state in which this parasitoid was collected on this host was in the Province of Missiones, Argentina (Schliserman *et al.*, 2010).

The fact that the guava tree fruits from March until May and June (Hoffmann *et al.*, 1994; Paise & Vieira, 2005) should be taken into account in biological control programs, as this fruit could act as a reservoir of parasitoids at a time when few fruits are found and temperatures commonly drop.

The three species of native parasitoids, *D. areolatus, U. anastrephae* and *A. pelleranoi*, were observed on the araçâs and guavas, both in the control group and in the group exposed to *D. longicaudata* (Table 2). These species have been reported to occur on both guava and araçâs in RS and Bahia (Matrangolo *et al.*, 1998; Cruz *et al.*, 2011; Carvalho, 2005c; Gatelli, 2006).

The fruiting season of araçâs and guavas, between February and mid-April, is favorable for the development of fruit flies and their parasitoids. Especially in southern Brazil, temperatures are still high during this period (late summer and early fall), which guarantees the maintenance of larger populations of parasitoids and a reduction in those of their hosts before winter arrives.

It is very difficult to simply compare the parasitism rates found in the test batches of fruit in this study with those recorded by other authors in the same fruit species. A number of factors affect the results obtained in each situation, such as weather conditions, handling and age of the fruit trees, the surroundings of the sample areas, proximity to rural or urban areas, the reduction in habitat size caused by cutting down forests and/or replacing native fruit species with non-fruit species, the size of the samples taken, the population fluctuation of the pest and the control methods adopted, which makes it difficult to interpret the differences observed.

Even so, guavas showed the highest parasitism rates, indicating that this fruit is an important refuge for native parasitoids and could be a multiplier for populations of these natural enemies. The preservation of refuge areas for the development of fruit fly parasitoids has already been advocated, as has the planting of native myrtaceae around commercial fruit production areas, by Aguiar- Menezes & Menezes (1997) and Hickel (2002).

With the exception of the persimmons, all the batches of fruit that had contact with *D. iongicaudata* showed significantly higher parasitism rates than the control batches, as well as significantly lower infestation values (flies/fruit) (Table 1). The lower infestation may be due to the mortality of larvae in the fruit, since, according to Ashley *et al.* (1976), batches of *A. suspensa* larvae exposed to this parasitoid showed around 33% higher mortality.

The fact that *D. longicaudata* has parasitized fruit flies in practically all fruit species shows that this parasitoid has great plasticity, being able to recognize a host even in a fruit that does not exist in its center of origin. This parasitoid attacks third instar tephritid larvae on fruit that has already fallen to the ground, often decomposing (Purcel *et al.*, 1994; Sivinski *et al.*, 2000; Sivinski *et al.*, 2001), in which the chemical cues may have been lost or reduced by being exposed to the weather. According to Thompson (1989) and Stuhl *et al.* (2011), the females *of D. longicaudata* are attracted to volatiles released by the fermentation of fallen fruit in the field, which explains the adaptation of this natural enemy to fruit of different origins.

According to Jang *et al.* (2000), it is wrong to isolate factors when studying the tritrophic interaction that occurs between the parasitoid, the fruit fly larva and the host fruit. When there is such a relationship, visual, tactile and olfactory stimuli complement each other. This means that the fruit is not the only variable when choosing an oviposition site.

Therefore, for biological control with an exotic species to be successful, it is important that this natural enemy can seek out its host in native fruit trees, since these are the main breeding grounds for fruit flies.

With regard to the fruit, the color, shape and stage of decomposition are factors that influence the search (Leyva *et al.*, 1991). In addition, the infestation of the fruit (Eben *et al., 2000) and the* origin of the parasitoids can be determining factors in the success of the search and parasitism (Bautista & Harris, 1997). Therefore, regardless of the origin of the fruit used in this study, it was expected that if there was a fruit fly infestation, parasitism would occur, even if parasitism rates varied, which they did.

In araçâs, guavas and guavas-serranas, all the native parasitoid species that had been found in the control plot were also collected in the group exposed to the exotic (Table 2), and in the latter fruit, the species *A. pelleranoi* was also recorded. The inclusion of *D. iongicaudata* altered the frequency with which they occurred, but no deletion was found. These results corroborate those of Garcia-Medell *et al.* (2007), who evaluated the parasitism of *A. ludens* on guavas in cages and found no deletion among six species of braconids, including *D. longicaudata*.

Parasitism by *D. longicaudata,* without the deletion of native species, had already been reported in other studies on guavas and pitangas, such as Purcel *et al.* (1994) in Hawaii, Carvalho (2005c) in Recôncavo Baiano, Alvarenga *et al.* (2005) and Alvarenga *et al.* (2009) in Minas Gerais and Vargas *et al.* (2012)

in Tahiti. In the witness plots of loquats and guabirobas, only one individual of *D. areolatus* had been observed in each. However, in the exposed fruit group, this braconid was not found (Table 2). It is believed that the absence of this parasitoid is not due to deletion, as Alvarenga *et al.* (2005) concluded that *D. longicaudata* and *D. areolatus* do not compete for the same niche. In the same vein, Sivinski *et al.* (1998) had already reported that *D. areolatus* and *D. longicaudata* coexist because, although both are parasitoids of third instar larvae, *D. areolatus* can attack younger larvae under extreme conditions, such as low temperatures. Thus, the *D. areolatus* larva may already be advanced in size when *D. longicaudata* lays its egg in the same host. It is possible that the native parasitoid, when parasitizing at stages prior to the 3rd instar, will be able to win the competition with the exotic.

In loquats, after contact with *D. iongicaudata*, there was a significant decrease in infestation, which went from 4.7 ± 0.39 pupae/fruit to 2.4 ± 0.26 pupae/fruit in the treatment batch (Table 1). The smallest difference between the parasitism index in the treatment batch and the control batch was recorded in these fruits (Table 1). This may have been caused indirectly by the method used to collect the fruit. While the fruits of the other fruit trees were collected from the ground, those of the loquats were removed from the canopy, as most of them don't fall, but remain on the plant after they have finished ripening, rotting and remaining on the branches until close to the flowering stage. Therefore, as the loquats were removed from the plants, it is possible that some or most of the fly larvae had not developed to the third instar stage, which is ideal for parasitizing *D. iongicaudata*.

Another important aspect to consider is the fact that native parasitoids and myrtaceae evolved in the same environment (Guimarâes *et al.*, 2004; Ovruski *et al.*, 2000). It is likely that the foraging and parasitizing abilities of these native organisms are not yet so well adapted to exotic fruit trees, such as loquats and peaches, which may help to explain why only one native specimen was recorded on loquats and none on peaches. In Mexico, as in the present study, there are no native parasitoids on peaches, and *D. longicaudata* is the only one that has been found (Sivinski *et al.*, 2000).

The increase in the parasitism index in peaches, although it has increased (Table 1), refers to only three individuals of *D. longicaudata*, out of a total of ten insects, seven of which were *A. fraterculus*. The small number of both flies and parasitoids shows that the phytosanitary treatments used on this fruit affected both groups of insects. Although no mortality of *D. longicaudata* was recorded when in contact with the fruit (all the females placed in the cage were alive 24 hours later), which could indicate the absence of toxic residues, the infestation on the peaches was low, leaving few hosts for the parasitoids.

The *A. fraterculus* infestation observed in the control batch of peaches was 0.44 ± 0.09 flies/fruit, much higher than that of the treatment group, which was 0.14 ± 0.05 flies/fruit. As the pressure of parasitism inside the cages was high and the *D. longicaudata* females used in the experiment were at the age at which they have maximum reproductive capacity, between five and seven days, it is possible that some larvae died.

Parasitism pressure is a determining factor in the increase in parasitism rates. In the present study, the indices reached values of between 14.75% in loquats and 68.4% in araçâs, but, as this was an experiment in cages, it is likely that this pressure was much higher than that possibly found in the field. For example, BomFim *et al.* (2010) recorded an increase in the parasitism index of only 5% in umbu-cajà fruit (*Spondias* spp.), in the field, 48 hours after the release of *D. longicaudata* in the Recôncavo Baiano region.

The peaches used in this study (Item 3.3), compared to the other fruit species, were large and had thick flesh. According to Leyva *et al.* (1991), in fruit with thicker flesh, parasitism rates tend to be lower, as the fruit fly larvae can go deeper, with few hosts remaining near the surface of the fruit. In the presence of few hosts, *D. longicaudata* tends to over-parasitize (Altafini *et al.*, 2013). The relationship between fruit size and superparasitism has already been evaluated by Montoya *et al.* (2013), who concluded that larger fruits tend to present more cases of superparasitism in fruit fly larvae.

According to Paladino *et al.* (2010), there is no emergence when *D. longicaudata* overparasitizes, as larval mortality occurs due to cannibalism, physical attacks, lack of food and/or oxygen. It is therefore considered that with a low supply of hosts, parasitoids tend to over-parasitize, reducing the emergence of flies, but leaving no offspring, from the point of view of longer-lasting biological control, which would not be a desirable strategy.

The pupae in which no emergence had been recorded, collected from all species of fruit, were dissected, but they were dry and it was not possible to identify the insect inside. In order to assess the occurrence of superparasitism, it would be necessary to dissect the larvae inside the fruit or the newly formed pupae, which was not done in this study.

The hypothesis that some larvae may have been killed by the parasitoids without completing their development is also supported by the viability of the pupal stage in the two batches of peaches. While the control batch produced 30 pupae, with 22 flies emerging, the exposed batch produced 23 pupae, with seven flies and three parasitoids emerging. Thus, 73.3% of the pupae were viable in the control, while 43.4% were viable in the treatment. Considering that the fruit was collected in the same place and at the same time and that the environmental conditions for all the fruit were the same, fly mortality may have influenced, if not as a whole, at least in part, the viability values of the pupal stage that were recorded in the plots exposed to *D. longicaudata*.

Superparasitism is a factor in fruit fly mortality and occurs naturally in parasitism by *D. longicaudata*, both in field and laboratory situations (Montoya *et al.*, 2000; Meirelles *et al.*, 2013; Montoya *et al*, 2013). Therefore, it was expected that larval mortality would be higher in the batch of fruit exposed to the parasitoid. This was a situation of high parasitism pressure, as the insects were in a restricted area, with ten other females competing for an unlimited number of hosts. According to Montoya *et al.* (2000), the greater the pressure of parasitism, the greater the number of occurrences of superparasitism. Meirelles *et al.* (2013) also stated that the period between five and ten days of life for this species is when the greatest number of cases of multiple eggs within the same host occur, an age that was used in this experiment.

The possibility of larval death can be raised for all the fruits used in this experiment exposed to *D. longicaudata*, as the number of pupae was always lower (Table 1), although only in loquats, araçâs and pitangas was the difference significant.

The results of this study show how difficult it is for native parasitoids to parasitize their hosts on exotic fruits. According to Hickel (2002), both parasitoids and fruit flies have similar host-seeking mechanisms. Therefore, it is normal for parasitism rates to follow the trend of infestation levels, being higher in native fruits. Therefore, the introduction of *D. longicaudata*, even under laboratory conditions, represented an important increase in parasitism on exotic fruit, which was practically non-existent before exposure to the exotic parasitoid.

Apparently, through the way the study was conducted, there was no deletion of natives when *D. iongicaudata* was introduced, but this conclusion can only be properly made after sampling over a period of years in the field.

The low frequency of native parasitoids found in all the fruit trees studied seems to highlight the scarcity of these organisms, which may be related to climatic conditions, the gradual destruction of habitats, deforestation, pesticide applications or even because the populations of these organisms may be naturally small.

Removing fruit from the field may underestimate parasitism, as some species need fruit in *an* advanced stage of decomposition or fruit that is no longer whole, such as *U. anastrephae* and *A. pelleranoi* (Alvarenga *et al.*, 2005). In this study, only whole fruits were collected, so it is possible that the actual parasitism rate is higher.

Another aspect to note is that parasitoid species have morphological differences that are related to oviposition preferences, such as host development stage, fruit size and/or skin thickness, which influence competition for oviposition sites (Sivinski *et al.,* 1997). In this way, the introduction of an exotic parasitoid that has different preferences and needs from the native ones can help to increase parasitism rates without causing a major environmental impact.

The increase in parasitism rates without the deletion of native parasitoids can be explained by the fact that species which prefer 3rd instar larvae are able to identify larvae which have already been parasitized, selecting only those hosts which guarantee the greatest probability of success for their offspring. Wang & Messing (2003) found that *Diachasmimorpha tryoni* (Cameron), which prefers 3rd instar larvae, did not compete with *F. arisanus*, which parasitizes 1st instar larvae. The authors concluded that parasitoids that attack different instars complement each other. However, according to the authors, *D. tryoni* is harmed when *F arisanus* parasitizes the eggs first, because when the eggs of the first species are laid, the latter can destroy or even feed on the eggs of the first.

In addition to the reasons already listed in the study by Sivinski *et al.* (1998), in Rio Grande do Sul, *D. areolatus* becomes competitive in relation to *D. longicaudata*, because it is more adapted to the climatic conditions, mainly because the exotic has a lower thermal threshold of 12.5 °C in *A. fraterculus*, which is the most abundant host, considering that often, in winter, the minimum is lower than this temperature (FEPAGRO, 2014).

The possibility of *D. longicaudata* causing the deletion of native species or greatly reducing their populations is unlikely given that, in some areas of Hawaii, for example, *D. longicaudata* was only the fourth most captured braconid, representing only 5.6% of parasitoids (Bokonon-Ganta *et al.*, 2007). This was because in the area where the study was carried out there were species that attacked earlier instars, including eggs. A similar result was reported by Vargas *et al.* (2012) in Tahiti, where *D. longicaudata* was responsible for increases of up to 5.5% in parasitism rates. According to the authors, at the study site, *D. longicaudata* was mainly harmed by *F arisanus*, which parasitizes eggs. These studies present results that reinforce the conjecture raised in this study, that *D. longicaudata* does not compete, or only competes to a low degree, with other parasitoid species.

Thus, based on the results obtained in this experiment, *D. longicaudata* has the potential to parasitize fruit flies in Rio Grande do Sul, without causing a major impact on native parasitoid populations, even in situations of intense parasitism pressure.

### 4.2 Parasitism *of Diachasmimorpha longicaudata* in semi-field conditions

Infestation, pupae/fruit, was significantly lower in loquats (t = 3.6903; gl = 46; P = 0.0006) and guabirobas (t = 2.0921; gl = 37.51; P = 0.0433) in plots exposed to *D. longicaudata*. There were no differences in guavas (U = 200.5; Z(U) = 1.8042; P = 0.0712), *kumquats* (t = 1.6253; gl = 46; P = 0.1108), araçâs (t = - 0.4204; gl = 46; P = 0.6761) and peaches (t = 1.9626; gl = 39.11; P = 0.0568) (Table 3).

TABLE 3 Average number of pupae/fruit and flies/fruit in the control batch (TE) (without exposure to *Diachasmimorpha longicaudata*) and treatment batch (TR) (with exposure to *D. longicaudata*) of *loquats*, guavas, guabirobas, *kumquats*, araçâs and peaches, in a semi-field test (n = 24 fruits of each species), Porto Alegre, RS.

| | **Puparium/fruit** | | **Fly/fruit** | |
|---|---|---|---|---|
| **Fruit species** | **TE** | **TR** | **TE** | **TR** |
| **Loquat** | 6,3 ± 0,77 A* | 2,9 ± 0,53 B | 5,1 ± 0,60 a | 1,9 ± 0,46 b |

| | | | | |
|---|---|---|---|---|
| **Guava** | 6,3 ± 1,02 A | 5,9 ± 1,63A | 4,9 ± 0,93 a | 3,5 ± 1,01 b |
| **Guabiroba** | 1,8 ± 0,35 A | 1,0 ± 0,21 B | 1,1 ± 0,24 a | 0,3 ± 0,12 b |
| ***Kumquat*** | 2,0 ± 0,28 A | 1,3 ± 0,26A | 1,3 ± 0,24 a | 0,7 ± 0,16 b |
| **Araçà** | 2,9 ± 0,31 A | 2,0 ± 0,22A | 1,9 ± 0,20 a | 0,7 ± 0,12 b |
| **Peach** | 3,2 ± 0,54 A | 2,0 ± 0,35A | 2,3 ± 0,37 a | 1,1 ± 0,19 b |

* Values followed by different letters differ from each other in the row (upper case for pupa/fruit and lower case for fly/fruit) using the t-test (α = 0.05) for *loquats*, guabirobas, *kumquats*, araçâs and peaches and the Mann-Whitney test (α = 0.05) for guavas.

The lower number of pupae formed in the loquats and guabirobas exposed to parasitism may indicate that some larvae were parasitized, and possibly killed, without being able to complete their development to the pupal stage, as reported by Ashley *et al.* (1976).

The number of flies/fruit was lower in all the fruit from the treatment lots (guabirobas: t = 2.8491; gl = 33.78; P = 0.0075; *kumquats* : t = 2.1582; gl = 46; P = 0.0361; araçà: t = 4.8426; gl = 37.8; P < 0.0001; loquat: t = 4.1782; gl = 46; P = 0.0001; peach: t = 2.722; gl = 35.39; P = 0.01; guava: U = 186; Z(U) = 2.1032; P = 0.0354) (Table 3), which can be explained by parasitism and/or mortality of the larvae (inviability).

As mentioned in section 4.1, a limited number of hosts can be responsible for an increase in cases of overparasitism (Montoya *et al.,* 2013), and this tends to be a factor in the mortality of fruit fly larvae in batches of fruit exposed to the parasitoid. As observed by Meirelles *et al.* (2013), the maximum reproductive performance of *D. longicaudata*, i.e. the greatest possibility of overparasitism, occurs between the fifth and tenth day of age, which was the period of development of the insects used in the experiment.

Controlled infestation by placing fruit flies in the bags surrounding the fruit ensured that the fruit came into contact with the insects, partly explaining the higher number of pupae per fruit compared to the experiment described in item 4.1 in the control plots. When infestation occurs naturally in the field, the density of flies per fruit is lower, resulting in a lower number of individuals per fruit.

Parasitism could be observed in both the laboratory (Tables 1 and 2) and semi-field (Table 4) experiments, which led to the mortality of fly larvae, reducing the number of pupae collected after contact with parasitoids. In the semi-field work, only *D. longicaudata* had contact with the fruit, while in the experiment described in item 4.1, the fruit had already had contact with native parasitoids in the field, which could have increased mortality, reducing the infestation observed in the control plot.

The females *of D. longicaudata* in the semi-field test were able to recognize and parasitize the hosts in native and exotic fruits, however, the number of offspring left per day was lower than that verified by Meirelles *et al.* (2013) in the laboratory, which recorded six to ten offspring per day. In the present study, if we consider the number of parasitoids that emerged from the pupae from the treatment batches, we can see that the number of offspring generated per female of *D. longicaudata* introduced into the cage was between 0.5 in *kumquats* and 1.33 in guavas (Table 4).

TABLE 4 Parasitism (%), number of parasitoids per fruit and offspring emerged per female in *loquats*, guavas, guabirobas, *kumquats*, araçâs and peaches from the treatment plot, Porto Alegre, RS.

| **Fruit species** | **Parasitism (%)** | **Parasitoid/fruit** | **Offspring/female** |
|---|---|---|---|
| **Loquat** | 20,3 | 0,50 | 1,00 |
| **Guava** | 16,0 | 0,66 | 1,33 |
| **Guabiroba** | 47,0 | 0,33 | 0,66 |
| ***Kumquat*** | 26,1 | 0,25 | 0,50 |
| **Araça** | 38,6 | 0,50 | 1,00 |
| **Peach** | 20,0 | 0,29 | 0,58 |

Some factors may explain the differences between the results of these two studies. Firstly, the insects used by Meirelles *et al.* (2013) were kept in controlled conditions throughout the experiment, while in the present study the cages were kept in a greenhouse and were therefore exposed to variations in temperature, humidity and solar radiation (Table 5), which can influence parasitism. According to Andrewartha & Birch (1954), temperature can affect not only the biology but also the fecundity and behavior of insects.

Another factor that may have influenced the results is that in the semi-field test, the fruit was kept in a larger environment, without controlled conditions, in a simulation closer to a natural environment. In this way, the fruit's volatiles had to be recognized by the females in an environment unknown to them. In the laboratory, parasitism took place on artificial substrates, only with fruit fly larvae, with few sources of volatiles or signals to find the host. In the cages in the greenhouse, the females had to search for the larvae by detecting the volatiles in a new environment, with surrounding vegetation, air pollution and other fruit on trees nearby. All the volatiles, as yet unknown to the female *D.*

*longicaudata* used in the experiment had to be recognized and selected. However, the fact that parasitism occurred under these conditions is considered a positive result, as females produced in the laboratory can be used for field releases.

TABLE 5 Meteorological data, minimum and maximum temperature (°C), sunshine (hours), average relative humidity (%) and rainfall (mm), during the days the fruit was exposed to the parasitoid *Diachasmimorpha longicaudata* in a semi-field test, Porto Alegre, RS, September 4, 2012 to February 28, 2013.

| **Fruit** | **Dates** | **Max temp (°C)** | **Min temp (°C)** | **Insulation (hours)** | **UR (%)** | **Precipitation (mm)** |
|---|---|---|---|---|---|---|
| | 04/Sep/12 | 26,1 | 18,6 | 8,7 | 64,5 | 1,4 |
| Loquat | 05/Sep/12 | 27,1 | 14,5 | 11,2 | 59,7 | 0 |
| | 06/Sep/12 | 28,0 | 15,0 | 10,8 | 66,5 | 0 |
| | 15/Oct/12 | 29,2 | 14,7 | 5,4 | 64,2 | 0 |
| Peach | 16/Oct/12 | 29,9 | 17,9 | 4,9 | 67,5 | 0 |
| | 17/Oct/12 | 25,7 | 18,4 | 10,8 | 57,2 | 0 |
| | 13/Nov/12 | 24,1 | 14,2 | 5,6 | 69,5 | 6,2 |
| Kumquat | 14/Nov/12 | 25,3 | 15,2 | 9,9 | 65,0 | 0 |
| | 15/Nov/12 | 27,1 | 16,6 | 11,9 | 71,5 | 0,1 |
| | 05/Feb/13 | 27,7 | 17,2 | 10,2 | 69,7 | 0 |
| Araçà | 06/Feb/13 | 29,6 | 19,9 | 11,3 | 71,2 | 0 |

|  |  |  |  |  |  |  |
|---|---|---|---|---|---|---|
|  | 07/Feb/13 | 29,5 | 20,8 | 12,0 | 66,7 | 0 |
|  | 26/Feb/13 | 26,1 | 18,6 | 8,7 | 64,5 | 1,4 |
| Guava | 27/Feb/13 | 27,1 | 14,5 | 11,2 | 59,7 | 0 |
|  | 28/Feb/13 | 28,0 | 15,0 | 11,3 | 66,5 | 0 |

Source: INMET, 8th district, Porto Alegre, RS (2014).

The parasitoids were able to seek out their hosts in the fruit, even though they were reared in artificial rearing units. According to Bautista & Harris (1997), *D. iongicaudata*, when reared artificially, does not stop looking for its hosts in fruit, but parasitism success is lower. The authors compared female parasitoids from *B. dorsalis* larvae reared on papaya and from larvae on an artificial diet. In the study, those that had papaya as a base for rearing the host and as a substrate for oviposition parasitized 30 to 40% more, reaching more than 60% success. However, when offered a different substrate to the one on which they had been reared, the females on the artificial diet parasitized up to 25% more, but only reached 40% success. It would therefore be interesting to carry out a test in which the females were given a period of experience with natural fruit, to see if their parasitism success would increase. The efficiency of the exotic parasitoid in the field may increase as it spends more time in the orchard environment and the surrounding forest.

The comparison of parasitism rates between the fruits could not be effectively evaluated. As the parasitoids did not have the chance to choose between the fruits, having only one species at their disposal, it cannot be said that there is a preference for one species or another. When you isolate fruits or factors to evaluate a tritrophic interaction, you run the risk of making a mistake when comparing the performance of the parasitoids (Jang *et al.*, 2000). For each fruit species, parasitism behavior can be different, depending on how it is perceived through volatiles, shape and color (Sthul *et al.*, 2011), as well as the thickness of the pulp and skin (Hickel 2002).

Although collecting fallen fruit from orchards has been a recommended practice for more than 50 years (Lipp & Secchi, 2002), producers don't usually do it. Therefore, the hosts for *D. iongicaudata* are not only available in the woods surrounding the orchards, they can be both inside and outside the cultivation sites. These environments differ in terms of floristic composition and weather conditions. Therefore, the ability of the females to parasitize fruit fly larvae immediately after being transferred from the farm to a semi-field situation is a very important factor.

The results of this work showed that, even after being reared in a laboratory, the females were able to seek out their hosts and oviposit on fruit in a more complex system than they had before being released into the cages of the semi-field test.

### 4.3 Infestation and parasitism in the field experiment

The predominance of *A. fraterculus* in all the fruit trees (Table 6) was expected, as it is a native species and therefore adapted to the environment. This result corroborates those of Gattelli *et al.* (2008), in Vale do Cai (RS) and Nunes *et al.* (2012), in Pelotas (RS).

The smallest infestations were found on persimmon and only on this fruit was *C. capitata* predominant, suggesting that *A. fraterculus* is less adapted to this host. Another hypothesis is that, because the persimmon tree fructifies between the end of March and the beginning of May, it competes as a host for *A. fraterculus*, temporally, with other native fruit trees that evolved together with the South American fruit fly, such as araçâs, which were present in the study area and must be more attractive.

Specimens of Lonchaeidae were collected from loquats, peaches and araçâs. Regarding the insects of this family, there is some disagreement about their ability to cause damage. In fact, they are carpophagous flies and feed on the fruit pulp, just like the tephritids. Some authors defend the idea that some species can become primary pests, causing damage to healthy fruit (Uchôa-Fernandes *et al.,* 2003; Raga *et al.*, 2004; Strikis & Prado, 2005), while others suggest that lonqueids oviposit only in holes left by tephritids or in wounds in the fruit skin (Souza *et al*., 1983; Malavasi *et al*., 1994).

The intensity of infestation (pupae/kg of fruit) was similar in the three years in peaches ($\chi^2$ = 3.381, gl = 2, P = 0.1844), apricots ($\chi^2$ = 2.094, gl = 2, P = 0.351) and persimmons (χ2 = 0.227, gl = 2, P = 0.8929). In loquats, infestation was statistically lower in the post-release year (χ2 = 6.848, gl = 2, P = 0.0326) (Table 6).

TABLE 6. Number of fruits, fruit weight, number of pupae, infestation (pupae/kg of fruit), number of flies, fly viability, parasitism index and parasitoid species on loquats, peaches, araçâs and persimmons, in the pre-release (Year 1) (2011/2012), release (Year 2) (2012/2013) and post-release (Year 3) (2013/2014) years of *Diachasmimorpha longicaudata* at the Agronòmica Experimental Station (30°06'16"S and 51°39'58"O), in Eldorado do Sul, RS.

| | **Loquat** | | | **Peach** | | |
|---|---|---|---|---|---|---|
| | **Years** | | | **Years** | | |
| | **1** | **2** | **3** | **1** | **2** | **3** |
| **Fruits (n°)** | 893 | 1.971 | 1.505 | 692 | 762 | 555 |
| **Weight (kg)** | 22,17 | 41,93 | 28,11 | 48,81 | 66,15 | 32,95 |
| **Pupils (n°)** | 2.024 | 3.762 | 1.424 | 389 | 484 | 167 |
| **Puparia/kg** | 91,2 A* | 89,7A | 50,7B | 7,9$^{NS}$ | 7,3 | 5,1 |
| **Flies found (n°)** | 1.192 | 2.24 | 962 | 261 | 235 | 94 |
| *Anastrepha fraterculus* | 1.187 | 2.163 | 939 | 241 | 205 | 89 |
| *Ceratitis capitata* | 0 | 30 | 20 | 17 | 17 | 2 |
| Lonchaeidae | 5 | 47 | 3 | 3 | 13 | 3 |
| **Fly viability (%)** | 58,9 | 59,5 | 67,5 | 67,1 | 48,5 | 56,1 |
| **Parasitoids (n°)** | 8 | 221 | 12 | 4 | 50 | 3 |
| **parasitism index (%)** | 0,6 B | 9,2 A | 1,2 B | 1,5 B | 21,2 A | 3,2 B |
| | | **Araçà** | | | **Khaki** | |
| | | **Years** | | | **Years** | |
| | **1** | **2** | **3** | **1** | **2** | **3** |
| **Fruits (n°)** | 884 | 1.365 | 1.433 | 443 | 393 | 332 |
| **Weight (kg)** | 4,19 | 9,14 | 7,79 | 51,58 | 38,73 | 35,89 |
| **Pupils (n°)** | 1.488 | 7.747 | 5.159 | 18 | 15 | 15 |
| **Puparia/kg** | 355,5$^{NS}$ | 847,7 | 662,7 | 0,34$^{NS}$ | 0,38 | 0,41 |
| **Flies found (n°)** | 751 | 5.323 | 2.874 | 5 | 6 | 5 |

| | | | | | | |
|---|---|---|---|---|---|---|
| *Anastrepha fraterculus* | 750 | 5.322 | 2.872 | 0 | 1 | 0 |
| *Ceratitis capitata* | 0 | 0 | 0 | 5 | 5 | 5 |
| Lonchaeidae | 1 | 1 | 2 | 0 | 0 | 0 |
| **Fly viability (%)** | 50,4 | 68,7 | 55,6 | 27,7 | 40 | 33,3 |
| **Parasitoids (n°)** | 18 | 969 | 107 | 0 | 0 | 0 |
| **parasitism index (%)** | 2,4 B | 15,4 A | 3,6 B | 0 | 0 | 0 |

The small number of pupae collected from peaches, when compared to those from loquats and araçâs, may be related to the toxic baits sprayed weekly to reduce the number of adults in the orchard. Thus, by reducing adult populations, the supply of larvae inside the fruit may also be reduced. Harter *et al.* (2010), evaluating peach orchards in the Pelotas region, found that fruit fly populations were significantly lower in orchards with applications of toxic baits (hydrolyzed protein + malathion) than those observed in untreated orchards and statistically equal to those that received insecticides as a top dressing, according to the producers' own criteria.

In addition to the effect of toxic baits, variations in fruit fly populations can occur due to factors related to the microclimate and the structure of the environment in the same region. In a study carried out in two municipalities in northern Argentina, for example, Schliserman *et al.* (2010) found very different infestations in peach orchards not treated with insecticides, with 8.55 pupae/kg of fruit in Laharrague and 45.33 pupae/kg of fruit in Montecarlo.

No parasitism was recorded in caquize trees, which may have been because this host was not recognized by the parasitoids or because of the low infestation observed in this fruit tree.

The highest parasitism rates in the pre- and post-release years were recorded in the araçazeiro. This fruit tree also had the highest richness and abundance of native parasitoids, as well as being the only one where specimens of *D. brasiliensis* were found (Table 7). An important aspect to consider is the fact that the native parasitoids and the fruit trees of the Myrtaceae family evolved in the same environment (Ovruski *et al.*, 2000; Guimarâes *et al.*, 2004). It is likely that the foraging and parasitizing abilities of these native organisms are not yet so well adapted to exotic fruit trees. These results corroborate those of Salles (1996) and Nunes *et al.* (2012), who observed higher rates of parasitism, abundance and richness of native parasitoids in Myrtaceae species, including the collection of araçâs, compared to exotic fruit trees. Schliserman *et al.* (2010) pointed out that species from this family are very important in northeastern Argentina for the multiplication and maintenance of fruit fly parasitoid populations.

TABLE 7. Total number of parasitoids, parasitism indices and relative frequency (%) of parasitoids, on three species of fruit, in the years of pre-release (2011/2012) (Year 1), release (2012/2013) (Year 2), and post-release (Year 3) of *Diachasmimorpha longicaudata*, at the Agronòmica Experimental Station (30°06'16"S and 51°39'58"O), in Eldorado do Sul, RS.

| | **Loquat** | | | **Peach** | | | **Araçà** | | |
|---|---|---|---|---|---|---|---|---|---|
| | **Years** | | | **Years** | | | **Years** | | |
| | **1** | **2** | **3** | **1** | **2** | **3** | **1** | **2** | **3** |
| **Total Parasitoids** | 8 | 221 | 12 | 4 | 50 | 3 | 18 | 969 | 107 |
| **parasitism index (%)** | 0,6 | 9,2 | 1,2 | 1,5 | 21,2 | 3,2 | 2,4 | 15,4 | 3,6 |
| **Frequency of each species (%)** | | | | | | | | | |
| *Diachasmimorpha longicaudata* | 0 | 99,1 | 0 | 0 | 94 | 0 | 0 | 88,5 | 0 |
| *Doryctobracon areolatus* | 100 | 0,9 | 50 | 0 | 0 | 66,6 | 11,1 | 1,5 | 31,8 |
| *Doryctobracon brasiliensis* | 0 | 0 | 0 | 0 | 0 | 0 | 5,6 | 0 | 1,9 |
| *Utetes anastrephae* | 0 | 0 | 50 | 0 | 0 | 33,3 | 27,8 | 9,9 | 31,8 |
| *Aganaspis pelleranoi* | 0 | 0 | 0 | 100 | 6 | 0 | 55,6 | 0 | 34,6 |

Given that the species of native parasitoids that were recorded in the pre-release year were also verified in the release and post-release years, it is considered that these were not deleted. This indicates that the introduction of *D. longicaudata* to the study site did not harm the parasitoids that already occurred naturally in that environment, confirming the results obtained in Bahia (Carvalho *et al.*, 2005b; Bomfim *et al.*, 2010) and Minas Gerais (Alvarenga *et al.*, 2005), after the release of *D. longicaudata*.

Native species that compete for the same resource, in this case fruit fly larvae, specialize in order to act and search for their host in different places on the plant and exploit different phases of their host's cycle (Sivinski *et al.*, 1997), in a process of divergence.

Thus, it is possible that the introduction of *D. longicaudata* did not cause damage to native parasitoid populations because each species has a specific niche or competition strategy. Sivinksi *et al.* (1998) reported that parasitoids respond to various environmental factors according to intrinsic species characteristics, which can influence interspecific competition. According to the authors, both aspects of the parasitoid's morphology and those relating to the fruit are responsible for the parasitism strategy and niche of each species.

The species recorded in this study, *D. areolatus, U. anastrephae* and *A. pelleranoi*, have characteristics that give them advantages in parasitism, or possibilities of competing with each other. According to Sivinski *et al.* (2000), *A. pelleranoi* has the ability to search for its hosts in openings or cracks in fruit on the ground. Thus, this parasitoid is able to search for larvae inside the fruit, where the larvae go deeper, escaping parasitism by other species.

*Utetes anastrephae* has, in the larval stage, larger and stronger jaws than other parasitoids and a robust body, so it is able to compete and eliminate larvae of other species, such as *D. areolatus*, inside the host's body (Aluja *et al.*, 2013).

As for *D. areolatus*, it usually parasitizes third instar larvae. However, in situations of competition, it can seek out hosts from earlier instars and parasitize in times of colder temperatures, which gives it advantages over other species, including *D. longicaudata* (Sivinski *et al.*, 1998).

According to Carvalho (2004), an important ecological aspect that makes it possible for *D. areolatus* and *U. anastrephae* to coexist with *D. longicaudata* is the fact that the first two species are able to seek their hosts in unripe fruit.

The parasitism indices were higher in the year of release, in loquats ($\chi^2$ = 138.307, gl = 2, P < 0.0001), peaches ($\chi^2$ = 40.278, gl = 2, P < 0.0001) and araçâs ($\chi^2$ = 293.044, gl = 2, P < 0.0001), which is due to the presence of *D.* longicaudata (Tables 6 and 7). *longicaudata* (Tables 6 and 7). The number of individuals of this species collected in the year in which the releases were made shows that the participation of this insect was a determining factor in the increase in parasitism rates (Table 7) and, overall, *D. longicaudata* was responsible for more than 88% of fruit fly parasitism. Increases in parasitism rates or reductions in fly populations have been seen in many of the places where *D. longicaudata* has been introduced, such as Florida (Baranowski *et al.*, 1993) and Mexico (Montoya *et al.*, 2000).

The time elapsed between releases and the recovery of *D. iongicaudata* individuals varied between the fruit trees (Table 8). When collections are made soon after releases, the parasitism potential of this species may be underestimated. According to Carvalho (2004), fruit collections after releases should take place after sufficient time for the fruit to be exposed to exotic parasitoids. The author states that collecting fruit on the plant or recently fallen to the ground may underestimate parasitism.

TABLE 8. Dates of first and last releases and first and last captures, number and average interval between releases of *Diachasmimorpha longicaudata* at the Agronòmica Experimental Station, UFRGS, Eldorado do Sul, RS (30°06'16"S and 51°39'58"O) (2012/2013).

| **Fruits** | **First release** | **No. of releases** | **Interval between releases (days)** | **Last release** | **First catch** | **Last catch** |
|---|---|---|---|---|---|---|
| **Loquat** | 07/08/2012 | 3 | 16,3 | 25/09/2012 | 23/09/2012 | 06/10/2012 |
| **Peach** | 19/10/2012 | 3 | 13,6 | 29/11/2012 | 19/11/2012 | 06/12/2012 |
| **Araça** | 12/02/2013 | 2 | 15 | 27/02/2013 | 07/02/2013 | 27/02/2013 |
| **Khaki** | 04/03/2013 | 2 | 16 | 20/03/2013 | - | - |

In the loquats, the first catch occurred one week after the first release (one individual) (Table 8), and 150 specimens were recovered fifteen days after the introduction.

No individuals of *D. longicaudata* were detected in the last sample of apricots in the year of release, on September 25, 2012. The same occurred with the first peach collection. It is important to consider the fact that the minimum temperatures between September 20 and 28, 2012 were below 10 °C (EEA-UFRGS, 2014). In the laboratory, Meirelles (2011) found that the base temperature for the development of *D. longicaudata* is 12.5 °C when reared on *A. fraterculus.* Therefore, it is possible that the population of the exotic parasitoid was affected by the temperatures recorded in the EEA in Eldorado do Sul during this period.

In peaches, it was only in the fourth collection, after two releases of this fruit, that 14 individuals of *D. iongicaudata* were found. Between the first release and the first record, seven days passed during which temperatures were below 12.5 °C, three of which were rained, a factor which may have partly hindered the parasitoid's survival and activity in the environment. The scarcity of hosts at this time is another explanation for what was observed, since no native fruit had been identified around the peach orchard.

According to Montoya & Cancino (2004), the release of parasitoids even at times when the pest population is low is important for maintaining the biological control agent. Continued releases should seek to increase the chances of encounters between *D. longicaudata* and its hosts. In this way, the results indicate that, in order to increase the parasitoid's participation in parasitism rates and reduce the fruit fly population, releases should be more frequent and constant than those carried out in this study. Montoya *et al.* (2000) stated that in order to maintain a high parasitoid population in the field, releases must not cease from one harvest to the next, but must be a constant activity throughout the duration of the biological control program.

After the first catch on peaches, specimens of *D. longicaudata* were recorded in all subsequent collections.

One of the interesting aspects of this study is that individuals of *D. longicaudata* were already captured in araçâs before the first release in this fruit (Table 8), indicating that this parasitoid was able to remain in the environment and migrate to locate the host in another fruit, considering that the time between the last collection in peaches and the first in araçà was 60 days.

With regard to the number and frequency of parasitoids to be released, it is conjectured that the complexity and heterogeneity of the environment where the biological control of fruit flies is to be carried out must be taken into account when calculating the number and frequency of parasitoids to be released. The EEA, in Eldorado do Sul, is a rural area with a few small clusters of houses in the surrounding area, no more than two kilometers away from the study area. Most of the properties are small (up to 5 ha), with pastures and grain crops, interspersed with small native fields. Thus, these aspects may be favorable to the exotic parasitoid, as they provide refuge and hosts throughout the year.

An important aspect to note is that *D. longicaudata* was able to identify and parasitize fruit fly larvae on both native and exotic hosts. This parasitoid is attracted to oviposition sites by volatiles from the decomposition process, as well as substances produced by fungi that can grow on fruit during rotting (Greany *et al.*, 1977). On the fruit, visual clues, such as oviposition marks or surfaces marked by the development of the fly larvae, help to identify the region where the host is located (Sthul *et al.*, 2011). To pinpoint the place where it should oviposit, the female of this species uses vibrations caused by the larva feeding inside the fruit (Lawrence, 1981). This indicates that *D. iongicaudata* is able to adapt to different species of fruit, as it uses generic clues, common to decomposing fruit or marks of oviposition activities or tephritid development in general.

According to Leyva *et al.* (1991), the fruit preferred by *D. iongicaudata* for oviposition in a given test under controlled conditions may not be the favorite in the field, where many variables interact (volatiles, skin thickness, previous experience, location on the plant, color and size of the fruit, for example), making the choice more complex.

*D. iongicaudata* was not recorded in any fruit tree in the year after release, indicating that this species did not make it through the winter. In the months of June, July and August, in RS, it is common for average daily temperatures to be below the lower thermal threshold of this insect (EEA-UFRGS, 2014), which would prevent the development of this parasitoid. In addition, the amount of fruit available at this time is low, which also makes it difficult to find hosts.

Although the parasitoid did not survive throughout the winter, its effect during the period when it was in the field at favorable temperatures may have been enough to reduce the initial population of fruit flies on loquats and peaches, according to the significantly lower record in the post-release year (Table 6). This indicates that, with continued releases at favorable times, *D. iongicaudata* could be an important biological control agent in fruit trees in southern Brazil.

## 5. CONCLUSION

Under the conditions in which the experiments were carried out, it can be concluded that:

- *Diachasmimorpha longicaudata* exposed to infested fruit, both in the laboratory and in the field, does not negatively affect native parasitoid populations;
- in the laboratory there is an increase in parasitism by *D. longicaudata*;
- in semi-field conditions, *D. iongicaudata* increases the mortality of *A. fraterculus* pupae;
- releases of *D. longicaudata* increase parasitism rates in loquats, peaches and apricots;
- native parasitoids recognize hosts in exotic fruits such as peaches and apricots in the field;
- *D. longicaudata* parasitizes fruit flies on both native and exotic fruits, in the laboratory and in the field;
- *D. longicaudata* does not survive the low temperatures recorded in winter at the UFRGS Agronomic Experimental Station in Eldorado do Sul, RS.

## 6. FINAL CONSIDERATIONS

The complexity of the fruit fly/parasitoid system must be better understood before a classical biological control program can be planned and implemented. It is not just a question of knowing the relationship between the exotic agent and the natives, but also of knowing the possible hosts and non-target organisms.

When a biological control program with parasitoid releases is established, constant monitoring of the host population can indicate the need to increase or decrease releases of the exotic agent to control the pest, in order to cause the least possible impact on the native parasitoid guild.

Although there is the possibility of establishing native parasitoid farms in order to increase their abundance in the field, we still don't know whether their release would significantly increase parasitism levels. Releasing already established natural enemies can be a costly practice, without the expected response, to the point of making the biological control program unfeasible. For this reason, the relationship between all the possible agents involved in the fruit fly complex should be the subject of further study. In addition to the laboratory part of rearing and the field part of parasitoid releases, the economic bias involved in educating research teams, field teams and producers has to be considered. Likewise, cultural management, monitoring, the use of selective insecticides and other control methods must be studied so that a solid project with a low risk of failure can be carried out in fruit-growing regions.

# 7. BIBLIOGRAPHICAL REFERENCES

AGROFIT. Ministry of Agriculture, Livestock and Supply. **Information bank.**

Available at: <http://www.agricultura.gov.br/servicos-e-sistemas/sistemas/agrofit>. Accessed on: October 10, 2014.

AGUIAR-MENEZES, E. L.; FERRARA, F. A. A.; MENEZES, E. B. Fruit flies. In: CASSINO, P. C. R.; RODRIGUES, W. C. (Coord.) **Citricultura Fluminense**: principais pragas e seus inimigos naturais. Seropédica: Ed. Universidade Rural, 2004. p. 67-84.

AGUIAR-MENEZES, E. L.; MENEZES, E. B. Natural occurrence of parasitoids *of Anastrepha* spp. (Diptera: Tephritidae) in different host plants, in Itaguai (RJ), Brazil. **Biological Control**, Orlando, v. 8, p. 1-6, 1997.

ALTAFINI, D. L.; REDAELLI, L. R.; JAHNKE, S. M. Superparasitism of *Ceratitis capitata* and *Anastrepha fraterculus* (Diptera: Tephritidae) by *Diachasmimorpha longicaudata* (Hymenoptera: Braconidae). **Florida Entomologist**, Gainesville, v. 96, n. 2, p. 391-395, 2013.

ALUJA, M. et al. Fruit infesting tephritids (Dip.: Tephritidae) and associated parasitoids in Chiapas, Mexico. **Entomophaga,** Paris, v. 35, p. 39-48, 1990.

Note: when there are more than 3 authors, only the first is indicated, followed by the expression et al.

ALUJA, M. et al. Inter-specific competition and competition-free space i the tephritid parasitoids *Utetes anastrephae* and *Doryctobracon areolatus* (Hymenoptera: Braconidae: Opiinae).

**Ecological Entomology**, London, v. 38, n. 5, p. 485-496, 2013.

ALVARENGA, C. D. et al. Introduction and recovery of the exotic parasitoid *Diachasmimorpha longicaudata* (Ashmead) (Hymenoptera: Braconidae) in commercial guava orchards in northern Minas Gerais. **Neotropical Entomology**, Londrina, v. 34, n.1, p. 133-136, 2005.

ALVARENGA, C. D. et al. Fruit flies (Diptera: Tephritidae) and their parasitoids on host plants in three municipalities in the north of the state of Minas Gerais. **Arquivos do Instituto Biológico**, Sâo Paulo, v. 76, n. 2, p. 195-204, 2009.

ANDREWARTHA, H. G.; BIRCH, L. C. The innate capacity for increase in numbers. In: ANDREWARTHA, H. G.; BIRCH, L. C. (Ed.). **The distribution and abundance of animals.** Chivago: University of Chicago Press, 1954. p. 31-54.

ANVISA. Pesticide residues in food. **Revista Saùde Pùblica**, Sâo Paulo, v. 40, n. 2, p. 361-363, 2006.

ARAUJO, E. L.; ZUCCHI, R. A. Hosts and infestation levels of *Neosilba pendula* (Bezzi) (Diptera: Lonchaeidae) in the region of Mossoró/Assu, RN. **Arquivos do Instituto Biológico**, Sâo Paulo, v.

69, n. 2, p. 91-94, 2002a.

ARAUJO, E. L.; ZUCCHI, R. A. Parasitoids (Hymenoptera: Braconidae) of fruit flies (Diptera: Tephritidae) in the Mossoró/Assu region, state of Rio Grande do Norte. **Arquivos do Instituto Biológico**, Sâo Paulo,_v. 69, n. 2, p. 65-68, 2002b.

ASHLEY, T. R.; GREANY, P. D.; CHAMBERS, D. L. Adult emergence in *Biosteres* (*Opius*) *longicaudatus* and *Anastrepha suspense* in relation to the temperature and moisture concentration of the pupation medium. **Florida Entomologist**, Gainesville, v. 59, n. 4, p. 391-396, 1976.

AYRES, M. et al. **BioEstat 5.0:** statistical applications in the biological and medical sciences. Belèm: Sociedade Civil Mamirauà/MCT-CNPq, 2007. 324 p.

BARANOWSKI, R.; GLENN, H.; SIVINSKI, J. Biological control of the caribbean fruit fly (Diptera: Tephritidae). **Florida Entomologist**, Gainesville, v. 76, n. 2, p. 245-251, 1993.

BAUTISTA, R. C.; HARRIS, E. J. Effect of insectary rearing on host preference and oviposition behavior of the fruit fly parasitoid *Diachasmimorpha longicaudata*. **Entomologia Experimentalis et Applicata**, Dordrecht, v. 83, n. 2, p. 213-218, 1997.

BESS, H. A.; BOSCH, R. V. D.; HARAMOTO, F. H. Fruit fly parasites and their activities in Hawaii.

**Proceedings Hawaiian Entomological Society**, Honolulu, v. 17, n. 3, p. 367-378, 1961.

BOKONON-GANTA, A. H.; MCQUATE, G. T.; MESSING, R. H. Natural establishment of a parasitoid comples on *Bactrocera latifons* (Diptera: Tephritidae) in Hawaii. **Biological Control**, Orlando, v. 42, n. 3, p. 365-373, 2007.

BOMFIM, Z. V.; CARVALHO, R. da S.; CARVALHO, C. A. L de. Interspecific relationships between native fruit fly parasitoids and the exotic braconid *Diachasmimorpha longicaudata* on umbu-cajà fruit. **Ciência Rural**, Santa Maria, v. 40, n. 1, p. 77-82, 2010. Available at: <http://www.scielo.br/pdf/cr/2009nahead/a422cr1694.pdf>. Accessed on: June 5, 2013.

BOTTON, M. et al. New alternatives for monitoring and controlling *Anastrepha fraterculus* (Wied., 1830) (Diptera: Tephritidae) in temperate fruit growing. In: CONGRESSO BRASILEIRO DE FRUTICULTURA, 22., 2012, Bento Gonçalves, RS. **Proceedings**... Bento Gonçalves: [s.n.], 2012. 1 CD-ROM.

BOTTON, M.; SORIA, S. J.; HICKEL, E. R. Pests of the seedless grapevine. In: Embrapa. (Org.) **Seedless grapes cultivars BRS Clara, BRS Linda and BRS Morena**. Brasilia: Embrapa, 2005. p. 1-8. v. 1.

BRAZIL. Normative Instruction No. 38, of October 14, 1999. Establishes the list of regulated A1, A2 and non-quarantine pests to be observed by Brazil's phytosanitary defense system. **Diàrio Oficial [da] Repùblica Federativa do** Brasil, Poder Executivo, Brasilia, DF, 26 Oct. 1999. Section 1.

CANCINO, J.; MONTOYA, P. Advances and perspectives in the mass rearing of fruit fly parasitoids in Mexico. In: INTERNATIONAL SYMPOSIUM ON FRUIT FLIES OF ECONOMIC IMPORTANCE, 7., 2006, Salvador. **Proceedings...** Salvador: [s.n.], 2006. p. 133-142.

CARVALHO, R. da S. **Laboratory and field studies with the exotic parasitoid *Diachasmimorpha longicaudata* Ashmead (Hymenoptera: Braconidae) in Brazil**. 2003. 182 f. Thesis (Doctorate) - Institute of Biosciences, University of Sâo Paulo, Sâo Paulo, 2003.

CARVALHO, R. da S. **Impact of the introduction of the exotic parasitoid *Diachasmimorpha longicaudata* on the complex of native parasitoids of fruit flies (Tephritidae) in the Recôncavo Baiano**. Cruz das Almas, BA: Embrapa Mandioca e Fruticultura Tropical, 2004. 6 p. (Technical Communication, 98).

CARVALHO, R. da S. **Methodology for monitoring fruit fly populations in commercial orchards**. Cruz das Almas, BA: Embrapa Mandioca e Fruticultura Tropical, 2005a. 17 p. (Circular Tècnica, 75).

CARVALHO, R. da S. Evaluation of inoculative releases of the exotic parasitoid *Diachasmimorpha longicaudata* (Ashmead) (Hymenoptera: Braconidae) in a diversified orchard in Conceiçâo do Almeida, BA. **Neotropical Entomology**, Londrina, v. 34, n. 5, p. 799-805, 2005b.

CARVALHO, R. da S. **Asynchrony of emergence as a method for separating adults of *Ceratitis capitata* and the exotic parasitoid *Diachasmimorpha longicaudata* in low-cost artificial rearing**. Cruz das Almas, BA: Embrapa Mandioca e Fruticultura Tropical, 2005c. 4 p. (Technical Circular, 78).

CARVALHO, R. da S. **Biocontrol of fruit flies:** history, concepts and strategies. Cruz das Almas, BA: Embrapa Mandioca e Fruticultura Tropical, 2006. 5 p. (Technical Circular, 83).

CARVALHO, R. da S.; NASCIMENTO, A. S. Rearing and use of *Diachasmimorpha longicaudata* for biological control of fruit flies. In: PARRA, J. R. P et al. (Ed.). **Biological control in Brazil:** parasitoids and predators. Sâo Paulo: Manole, 2002. p. 65-179.

CARVALHO, R. da S.; NASCIMENTO, A. S.; MATRANGOLO, W. J. R. **Methodology for rearing the exotic parasitoid *Diachasmimorpha longicaudata* (Hymenoptera: Braconidae), for laboratory and field studies**. Cruz das Almas, BA: Embrapa Mandioca e Fruticultura Tropical, 1998. 16 p. (Technical Communication, 30).

CARVALHO, R. da S.; NASCIMENTO, A. S.; MATRANGOLO, W. J. R. Biological control. In: MALAVASI, A.; ZUCCHI, R. A. (Ed.). **Fruit flies of economic importance in Brazil:** basic and applied knowledge. Ribeirâo Preto: Holos, 2000. p. 113-117.

CLAUSEN, C. P. **Entomophagus insects**. New York: McGraw-Hill, 1940. 688 p.

CRUZ, P P; NEUTZLING, A. S.; GARCIA, F. R. M. First record of *Trichopria anastrephae*, a parasitoid of fruit flies, in Rio Grande do Sul. **Ciência Rural**, Santa Maria, v. 41, n. 8, p. 1297-1299, 2011.

DIAS, N. P. et al. Level of fruit fly infestation in a border strip in Rio Grande do Sul. **Revista Ceres**, Viçosa, v. 60, n. 4, p. 589-593, 2013.

EBEN, A. et al. Host species and host plant effects on preference and performance of *Diachasmimorpha longicaudata* (Hymenoptera: Braconidae). **Environmental Entomology**, Lanham, v. 29, n. 1, p. 87-94, 2000.

EEA-UFRGs - UFRGS Agronomic Experimental Station. **Meteorological data from the Agronomic Experimental Station - 1970-2012**. Available at:

<http://www.ufrgs.br/agronomia/joomla/index.php/eea-pesquisa>. Accessed on: 10 Dec. 2014.

FEPAGRO. **Meteorological Bulletin of Rio Grande do Sul, 2012**. Available at: <http://www.cemet.rs.gov.br/area/11/Produtos_e_Servi%C3%A7os>. Accessed on: December 10, 2014.

FOFONKA, L. Incidence and management of fruit flies (Diptera: Tephritidae) in orange orchards in the municipality of Caraà, Rio Grande do Sul. **Revista Brasileira de Agroecologia**, Porto Alegre, v. 2, n. 2, p. 834-837, 2007.

FUNASAKI, G. Y. et al. A review of biological control introductions in Hawaii: 1890 to 1985. **Proceedings Hawaiian Entomological Society**, Honolulu, v. 28, p. 105-160, 1988.

GATTELLI, T **Frugivorous flies (Diptera: Tephritidae) and parasitoids associated with myrtaceae and laranjeira "Céu" in Montenegro and Harmonia, RS**. 2006. 71 f. Dissertation (Master's Degree) - Postgraduate Program in Plant Science, Faculty of Agronomy, Federal University of Rio Grande do Sul, Porto Alegre, 2006.

GATTELLI, T. et al. Frugivorous flies associated with myrtaceae and laranjeira "Céu" in the Vale do Rio Cai region, Rio Grande do Sul, Brazil. **Ciência Rural**, Santa Maria, v. 38, n. 1, p. 236-239, 2008.

GARCIA-MEDEL, D. et al. Foraging behavior by six fruit fly parasitoids (Hymenoptera: Braconidae) released as single- or multiple-species cohorts in field cages: influence of fruit location and host density. **Biological Control**, Orlando, v. 43, p. 12-22, 2007.

GLEN, M. et al. *Puccinia psidii*: a threat to the Australian environment and economy - a review. **Australasian Plant Pathology**, Toowoomda East, v. 36, n. 1, p. 1-16, 2007.

GREANY, P. D. et al. Chemically mediated host finding by *Biosteres* (*Opius*) *longicaudatus*, a parasitoid of tephritid fruit fly larvae. **Journal of Chemical and Ecology**, New York, v. 3, n. 2, p. 189195, 1977.

GUILLEN, J. A.; ENKERLIN, D.; WONG, T T Y Population reduction of fruit flies *Anastrepha* spp. by inundative releases of *Diachasmimorpha longicaudata* (Hymenoptera: Braconidae) in Mazapa de Macadera, Chiapas, Mexico. **Vida Tephritidae**, México, v. 8, p. 5-6, 1991.

GUIMARAES, J. A. et al. Species of Eucoliinae (Hymenoptera: Cynipoidea: Figitidae) parasitoids of frugivore larvae (Diptera: Tephritidae and Lonchaiedae) in Brazil. **Anais da Sociedade Entomologica do Brasil**, Londrina, v. 28, n. 2, p. 263-273, 1999.

GUIMARAES, J. A. et al. Survey and tritrophic interactions of figitoids (Hymenoptera: Eucoilinae) parasitoids of frugivorous larvae (Diptera) in Brazil. **Arquivos do Instituto Biológico**, Sâo Paulo, v. 71, n. 1, p. 51-56, 2004.

HARRIS, E. J. et al. Suppression of melon fly (Diptera: Tephritidae) populations with releases of *Fopius arisanus* and *Psyttalia fletcheri* (Hymenoptera: Braconidae) in North Shore Oahu, HI, USA. **BioControl**, Dordrecht, v. 55, n. 5, p. 593-599, 2010.

HARTER, W. da R. et al. Toxic bait and sexual disruption in the control of South American fruit fly and oriental moth in peach trees. **Pesquisa Agropecuària Brasileira**, Brasilia, v. 45, n. 3, p. 229-235, 2010.

HENNEMAN, M. L. et al. Response to walnut olfactory and visual cues by the parasitic wasp *Diachasmimorpha juglandis.* **Journal of Chemical Ecology**, New York, v. 28, n. 11, p. 2221-2244, 2002.

HERNANDEZ, D. C. **Mediterranean fly.** Servicio Nacional de Sanidad Inocuidad y Calidad Agroalimentaria. Available at: <http://www.senasica.gob.mx/?id=4160>. Accessed on: 08 jul. 2014.

HICKEL, E. R. Pulp thickness as a conditioner of parasitism of fruit flies (Diptera: Tephritidae) by Hymenoptera: Braconidae. **Ciência Rural**, Santa Maria, v. 32, n. 6, p. 1005-1009, 2002.

HOFFMANN, A. et al. Influence of temperature and polyethylene on the storage of guava fruit (*Feijoa sellowiana* Berg.) **Scientia Agricola**, Piracicaba, v. 51, n. 3, p. 563-568, 1994.

INMET. **Meteorological Database for Teaching and Research**. 2014. Available at: <http://www.inmet.gov.br/portal/index.php?r=bdmep/bdmep>. Accessed on: October 28, 2014.

JANG, E. B. et al. Flight tunnel responses of *Diachasmimorpha longicaudata* (Ashmead) (Hymenoptera: Braconidae) to olfactory and visual stimuli. **Journal of Insect Behavior**, Berlin, v. 13, n. 4, p. 525-538, 2000.

KOVALESKI, A.; SUGAYAMA, R. L.; MALAVASI, A. Chemical control in apple trees. In: MALAVASI, A.; ZUCCHI, R. A. (Ed.). **Fruit flies of economic importance in Brazil:** basic and applied knowledge. Ribeirâo Preto: Holos, 2000. p. 135-141.

KOVALESKI, A.; RIBEIRO, L. G. Pest management in integrated apple production. In: PROTAS, J. F. S.; SANHUEZA, R. M. V. **Integrated fruit production:** the case of apples in Brazil. Bento Gonçalves: Embrapa Uva e Vinho, 2003. p. 61-76. (Technical Circular, 34).

LAWRENCE, P. O. Host vibration - a cue to the host location by the parasite *Biosteres longicaudatus.* **Oecologia**, Berlin, v. 48, p. 249-251, 1981.

LEYVA, J. L.; BROWNING, H. W.; GILSTRAP, F. E. Effect of host fruit species, size, and color on parasitization of *Anastrepha ludens* (Diptera: Tephritidae) by *Diachasmimorpha longicaudata* (Hymenoptera: Braconidae). **Environmental entomology**, Lanham, v. 20, n. 5, p. 1469-1474, 1991.

LIPP, J. P.; SECCHI, V. A. Fruit bagging: an old ecological practice for fruit fly control. **Agroecologia e Desenvolvimento Rural Sustentàvel**, Porto Alegre, v. 3, n. 4, p. 53-58, 2002.

LÓPEZ, M.; ALUJA, M.; SIVINSKI, J. Hymenopterous larval-pupal and pupal parasitoids of *Anastrepha* lies (Diptera: Tephritidae) in Mexico. **Biological Control**, Orlando, v. 15, n. 2, p. 119129, 1999.

MALAVASI, A.; NASCIMENTO, A. S; CARVALHO, R. da S. Fruit flies in IPM-Citrus. In: SEMINARIO INTERNACIONAL DE CITROS-MIP: Manejo integrado de Pragas, 3., 1994, Campinas. **Proceedings**... Campinas: Cargill Foundation, 1994. p. 211-231.

MARINHO, C. F. et al. Parasitoids (Hymenoptera: Braconidae) of fruit flies (Diptera: Tephritidae) in the State of São Paulo: associated plants and parasitism. **Neotropical Entomology**, Londrina, v. 38, n. 3, p. 321-326, 2009.

MATRANGOLO, W. J. R. et al. Parasitoids of fruit flies (Diptera: Tephritidae) associated with tropical fruit trees. **Anais da Sociedade Entomològica do Brasil,** Londrina, v. 27, n. 4, p. 593-603, 1998.

MEIRELLES, R. N. **Biology and thermal requirements of *Diachasmimorpha longicaudata* (Ashmead) (Hymenoptera: Braconidae) on *Anastrepha fraterculus* (Wiedemann) and *Ceratitis capitata* (Wiedemann) (Diptera: Tephritidae).** 2011. 67 f. Dissertation (Master's Degree) - Postgraduate Program in Plant Science, School of Agronomy, Federal University of Rio Grande do Sul, Porto Alegre, 2011.

MEIRELLES, R. N. et al. Parasitism of fruit flies (Thephritidae) on pitanga fruit in Harmonia, RS. In: SALAO DE INICIAÇÂO CIENTiFICA, 17., 2005, Porto Alegre. **Abstracts**... Porto Alegre: UFRGS, 2005.

MEIRELLES, R. N.; GATTELLI, T.; REDAELLI, L. R. Incidence and parasitism of fruit flies (Diptera: Tephritidae) on guava fruit in Vale do Cai, RS. In: SALÂO DE INICIAÇÂO CIENTiFICA, 18., 2006, Porto Alegre. **Abstracts**... Porto Alegre: UFRGS, 2006.

MEIRELLES, R. N.; REDAELLI, L. R.; OURIQUE, C. B. Comparative biology of *Diachasmimorpha longicaudata* (Hymenoptera: Braconidae) reared on *Anastrepha freterculus* and *Ceratitis capitata* (Diptera: Tephritidae). **Florida Entomologist**, Gainesville, v. 96, n. 2, p. 412-418, 2013.

MENDES, P. C. D. et al. Population assessment of fruit flies (Diptera: Tephritidae) and their larval parasitoids (Hymenoptera: Braconidae). **Revista Brasileira de Agroecologia**, Porto Alegre, v. 2, n. 1, p. 690-693, 2007.

MONTOYA, P.; CANCINO, J. Biological control by augmentation in fruit flies (Diptera: Tephritidae). **Folia Entomològica Mexicana**, México, v. 43, n. 3, p. 257-270, 2004.

MONTOYA, P. et al. The augmentative biological control component in the Mexican National Campaign Against *Anastrepha* spp. fruit flies. In: VREYSEN, M. J. B.; ROBINSON, A. S.;

HENDRICH, J. (Ed.). **Area-wide control of insect pests:** from research to field implementation. Netherlands: Springer, 2007. p. 661-670.

MONTOYA, P. et al. Biological control of *Anastrepha* spp. (Diptera: Tephritidae) in mango orchards through augmentative releaes *of Diachasmimorpha longicaudata* (Ashmead) (Hymenoptera: Braconidae). **Biological Control**, Orlando, v. 18, n. 3, p. 216-224, 2000.

MONTOYA, P. et al. Field superparasitism by *Diachasmimorpha longicaudata* attacking *Anastrepha* spp. larvae on mango fruits. **Biological Control**, Orlando, v. 64, n. 2, p. 160-165, 2013.

NASCIMENTO, A. S.; CARVALHO, R. da S. Integrated management of fruit flies In: MALAVASI, A.; ZUCCHI, R. A. (Ed.). **Fruit flies of economic importance in Brazil:** basic and applied knowledge. Ribeirâo Preto: Holos, 2000. p.169-173.

NAVA, D. E.; BOTTON, M. **Bioecology and control of *Anastrepha fraterculus* and *Ceratitis capitata* in peach trees**. Pelotas: Embrapa Clima Temperado, 2010. 29 p. (Documentos, 315).

NORA, I.; HICKEL, E. Apple tree pests: dipterans and lepidopterans. In: EPAGRI (Ed.). **Apple cultivation**. Florianópolis: GMC/EPAGRI, 2006. chap. 15, p. 463-486.

NUNES, A. M. et al. Frugivorous flies and their parasitoids in the municipalities of Pelotas and Capâo do Leâo, Rio Grande do Sul, Brazil. **Ciência Rural**, Santa Maria, v. 42, n. 1, p. 6-12, 2012.

OLIVEIRA, K. M. de; LUCCHESE, G. Sanitary control of pesticides in Brazil: the case of methamidophos. **Tempus Actas de Saùde Coletiva**, Brasilia, v. 7, n. 1, p. 211-224, 2013.

OLIVEIRA, M. R. V.; PAULA-MORAES, S. V. de. **Potential quarantine fruit flies for Brazil**. Brasilia: Embrapa Recursos Genéticos and Embrapa Cerrados, 2006. 261 p.

OVRUSKI, S. et al. Hymenopteran parasitoids on fruit-infesting Tephritidae (Diptera) in Latin America and the southern United States diversity, distribution, taxonomic status and their use in fruit fly biological control. **Integrated Pest Management Reviews**, Dordrecht, v. 5, p. 81-107, 2000

PAISE, G.; VIEIRA, E. M. Fruit production and spatial distribution of angiosperms with zoochorous fruits in a Mixed Ombrophilous Forest in Rio Grande do Sul, Brazil. **Revista Brasileira de Botànica**, Sâo Paulo, v. 28, n. 3, p. 615-625, 2005.

PALADINO, L. Z. C.; PAPESCHI, A. G.; CLADERA, J. L. Immature stages of development in the parasitoid wasp, *Diachasmimorpha longicaudata*. **Journal of Insect Science**, Oxford, v. 10, n. 56, 2010. Available at: <insectscience.org/10.56>. Accessed on: Oct. 2013.

Note: add access day.

PARANHOS, B. J. et al. Interspecific competition of parasitoids (Hymenoptera: Braconidae) of fruit flies (Diptera: Tephritidae). In: CONGRESSO BRASILEIRO DE FRUTICULTURA, 22, 2012, Bento

Gonçalves, RS. **Proceedings**... Bento Gonçalves: [s.n.], 2012. 1 CD-ROM.

PEREIRA-RÊGO, D. R. G. et al. Variation in fruit fly (Diptera: Tephritidae) infestation and parasitism at different fruiting stages in native myrtaceae in Rio Grande do Sul. **EntomoBrasilis**, Vassouras, v. 6, n. 2, p. 141-145, 2013.

PROTAS, J. F. da S.; VALDEBENITO-SANHUEZA, R. M. **Integrated fruit production**: the case of apples in Brazil. Bento Gonçalves: Embrapa Uva e Vinho, 2003. 192 p.

PURCELL, M. F. et al. Influence of guava ripening on parasitism of the oriental fruit-fly, *Bactrocera dorsalis* (Hendel)(Diptera: Tephritidae), by *Diachasmimorpha longicaudata*

(Ashmead)(Hymenoptera: Braconidae) and other parasitoids. **Biological Control**, Orlando, v. 4, n. 4, p. 396-403, 1994.

RAGA, A. et al. Fruit fly (Diptera: Tephritoidea) infestation in citrus in the State of Sâo Paulo, Brazil. **Neotropical Entomology**, Londrina, v. 33, n. 1, p. 85-89, 2004.

RAGA, A. et al. Host ranges and infestation indices of fruit flies (Tephritidae) and lance flies (Lonchaeidae) in Sao Paulo State, Brazil. **Florida Entomologist**, Gainesville, v. 94, n. 4, p. 787794, 2011.

SALLES, L. A. B. **Bioecology and control of the South American fruit fly.** Pelotas: Embrapa/CPACT, 1995. 58 p.

SALLES, L. A. B. Parasitism of *Anastrepha fraterculus* (Wied.) (Diptera: Tephritidae) by Hymenoptera, in the region of Pelotas, RS. **Pesquisa Agropecuâria Brasileira**, Brasilia, v. 31, n. 11, p. 769-774, 1996.

SALLES, L. A. B.; RECH, N. L. Effects of extracts of neem (*Azadirachta indica*) and cinnamon (*Melia azedarach*) on *Anastrepha fraterculus* (Wied.) (Diptera: Tephritidae). **Revista Brasileira de Agrociência**, Pelotas, v. 5, n. 3, p. 225-227, 1999.

SANTOS, J. P. et al. Incidence of white rot in apple fruits with and without wounds. **Revista Brasileira de Fruticultura**, Jaboticabal, v. 30, n. 1, p. 118-121,2008.

SCHLISERMAN, P.; OVRUSKI, S. M.; DE COLL, O. R. The establishment of *Diachasmimorpha longicaudata* (Hymenoptera: Braconidae) in Misiones, Northeastern Argentina. **Florida Entomologist,** Gainesville, v. 86, n. 4, p. 491-492, 2003.

SCHLISERMAN, P. et al. Diversity and abundance of hymenopterous parasitoids associated with *Anastrepha fraterculus* (Diptera: Tephritidae) in native and exotic host plants in Misiones, Northeastern Argentina. **Florida Entomologist**, Gainesville, v. 93, n. 2, p. 175-182, 2010.

SEGURA, D. F. et al. Response of the fruit fly parasitoid *Diachasmimorpha longicaudata* to host and host-habitat volatile cues. **Entomologia Experimentalis et Applicata**, Dordrecht, v. 143, n. 2, p. 164-176, 2012.

SILVA, F. F. da et al. Diversity of flies (Diptera: Tephritidae and Lonchaeidae) in organic citrus orchards in the Vale do Rio Cai, Rio Grande do Sul, Southern Brazil. **Neotropical Entomology**, Londrina, v. 35, n. 5, p. 666-670, 2006.

SILVA, F. F. da et al. Comparison of fruit fly control methods (Diptera: Tephritidae) in organic citrus production. **Revista da Faculdade de Zootecnia, Veterinària e Agronomia**, Uruguaiana, v. 14, n. 1, p. 36-52, 2007.

SIVINSKI, J. et al. Phenological comparison of two braconid parasitoids of the caribbean fruit fly (Diptera: Tephritidae). **Environmental Entomology**, Lanham, v. 27, n. 2, p. 360-365,1998.

SIVINSKI, J.; ALUJA, M.; LOPEZ, M. Spatial and temporal distributions of parasitoids of Mexican *Anastrepha* species (Diptera: Tephritidae) within the canopies of fruit trees. **Annals of the Entomological Society of America**, Oxford, v. 90, n. 5, p. 604-618, 1997.

SIVINSKI, J. M. et al. Suppression of a Caribbean fruit fly (*Anastrepha suspensa* (Loew) Diptera: Tephritidae) population through augmented releases of the parasitoid *Diachasmimorpha longicaudata* (Ashmead) (Hymenoptera: Braconidae). **Biological Control**, Orlando, v. 6, n. 2, p. 177-185, 1996.

SIVINSKI, J.; PINERO, J.; ALUJA, M. The distributions of parasitoids (Hymenoptera) of *Anastrepha* fruit flies (Diptera: Tephritidae) along an altitudinal gradient in Veracruz, Mexico. **Biological Control**, Orlando, v. 18, n. 3, p. 258-269, 2000.

SIVINSKI, J.; VULINEC, K.; ALUJA, M. Ovipositor length in a guild of parasitoids (Hymenoptera: Braconidae) attacking *Anastrepha* spp. fruit flies (Diptera: Tephritidae) in southern Mexico. **Annals of the Entomological Society of America**, Oxford, v. 94, n. 6, p. 886-895, 2001.

SOUZA, H. M. L. et al. Occurrence *of Anastrepha fraterculus* (Wied.), *Ceratitis capitata* (Wied.) (Diptera: Tephritidae) and *Silba* spp. (Diptera: Lonchaeidae). Eggs in oviposition bores on three host fruits. **Revista Brasileira de Entomologia**, Sâo Paulo,_v. 27, p. 191-195, 1983.

SOUZA, S. A. et al. Natural infestation of frugivorous flies (Diptera: Tephritoidea) in Arabica coffee, under wooded organic cultivation and in full sun, in Valença, RJ. **Neotropical Entomology**, Londrina, v. 34 n. 4, p. 639-648, 2005.

SOUZA FILHO, M. F. de et al. Diversity and seasonality of fruit flies (Diptera: Tephritidae and Lonchaeidae) and their parasitoids (Hymenoptera: Braconidae and Figitidae) in orchards of guava, loquat and peach. **Brazilian Journal of Biology**, Sâo Carlos, v. 69, n. 1, p. 31-40, 2009.

STRIKIS, P. C.; PRADO, A. P. A new species of the genus *Neosilba* (Diptera: Lonchaeidae). **Zootaxa**, Auckland, v. 828, n. 1, p. 1-4, 2005.

STUHL, C. et al. A compound produced by fruigivorous Tephritidae (Diptera) larvae promotes oviposition behavior by the biological control agent *Diachasmimorpha longicaudata* (Hymenoptera: Braconidae). **Environmental Entomology**, Lanham, v. 40, n. 3, p. 727-736, 2011.

TERAN, H. R. Feeding behavior and its correlation to reproduction in hembras *of Ceratitis capitata* (Wied.) (Diptera: Tephritidae). **Revista Agronómica N. O. A.,** Tucumàn, v. 14, p. 17-34, 1977.

THOMPSON, C. R. *Diachasmimorpha longicaudata* (Ashmead) (Hymenoptera: Braconidae), biological control agent for the Caribbean fruit fly. Department of Agriculture and Consumer Services, Division Plant Industry. **Entomology Circular**, Gainesville, n. 325, p. 325-326, 1989.

UCHÔA-FERNANDES, M. A. et al. Biodiversity of frugivorous flies (Diptera: Tephritoidea) captured in citrus groves, Mato Grosso do Sul, Brazil. **Neotropical Entomology**, Londrina, v. 32, n. 2, p. 239-246, 2003.

VALLE, G. E. do. **Biological and morphological aspects of *Diachasmimorpha longicaudata* (Ashmead, 1905) (Hymenoptera: Braconidae) reared on irradiated larvae of *Ceratitis capitata* (Wiedemann, 1824) (Diptera: Tephritidae)**. 2006. 85 f. Thesis (PhD) - Postgraduate Program in Sciences, Center for Nuclear Energy for Agriculture, University of Sâo Paulo, Piracicaba, 2006.

VARGAS, R. I. et al. Population dynamics of three *Bactrocera* spp. fruit flies (Diptera: Tephritidae) and two introduced natural enemies, *Fopius arisanus* (Sonan) and *Diachasmimorpha longicaudata* (Ashmead) (Hymenoptera: Braconidae), after an invasion by *Bactrocera dorsalis* (Hendel) in Tahiti. **Biological Control**, Orlando, v. 60, n. 2, p. 199-206, 2012.

WANG, X.; MESSING, R. H. Intra- and interspecific competition by *Fopius arisanus* and *Diachasmimorpha tryoni* (Hymenoptera: Braconidae) parasitoids of tephritid fruit flies. **Biological Control**, Orlando, v. 27, n. 3, p. 251-259, 2003.

WHARTON, R. A. Classical biological control of fruit-infesting Tephritidae. In: ROBINSON, A. S.; HOOPER, G. (Ed.). **World Crop Pests - Fruit Flies:** their biology, natural enemies and control. v. 3B. Elsevier: Netherlands, 1989. p. 303-313.

WHARTON, R. A. et al. Hymenopterous egg-pupal and larval-pupal parasitoids of *Ceratitis capitata* and *Anastrepha* spp.[Dip.: Tephritidae] in Costa Rica. **Entomophaga**, Paris, v. 26, n. 3, p. 285-290, 1981.

WHARTON, R. A.; YODER, M. J. **Parasitoids of fruit-infesting Tephritidae**. Available at: <http://paroffit.org/public/site/paroffit/home>. Accessed on: October 10, 2014.

ZUCCHI, R. A. Taxonomy. In: MALAVASI, A.; ZUCCHI, R. A. (Ed.). **Fruit flies of economic importance in Brazil:** basic and applied knowledge. Ribeirâo Preto: Holos, 2000. p. 13-24.

ZUCCHI, R. A. **Fruit flies in Brazil** - *Anastrepha* species and their hosts plants and parasitoids. Department of Entomology and Acarology, USP. 2008. Available at: <www.lea.esalq.usp.br/anastrepha/>. Accessed on: 10 Dec. 2014.

# 8. APPENDICES

APPENDIX 1. Average temperatures (°C), average minimum temperatures (°C), absolute minimum temperatures (°C), number of days with average temperatures below 10 °C, average maximum temperatures (°C), absolute maximum temperatures (°C), amount of rainfall (mm) and number of days with rainfall, monthly, from January 2011 to April 2014, at the UFRGS Agronomic Experimental Station, Eldorado do Sul, RS (30°06'16"S and 51°39'58"O).

| | 2011 | | | | | | | | | | | |
|---|---|---|---|---|---|---|---|---|---|---|---|---|
| | **January** | **February** | **March** | **April** | **May** | **June** | **July** | **August** | **September** | **October** | **November** | **December** |
| **Average Temperature (°C)** | 24,9 | 23,9 | 21,9 | 18,9 | 14,8 | 11,9 | 11,6 | 12,5 | 15,3 | 18,4 | 20,6 | 21,7 |
| **Average minima (°C)** | 20,7 | 19,7 | 17,2 | 13,3 | 9,4 | 7,0 | 6,8 | 8,1 | 9,1 | 12,9 | 13,4 | 15,8 |
| **Absolute min. (°C)** | 18,6 | 15,2 | 10,0 | 7,8 | 3,3 | -0,4 | -3,0 | 0,3 | 1,4 | 6,1 | 6,2 | 8,4 |
| **Days below 10°C** | 0 | 0 | 0 | 0 | 0 | 8 | 9 | 8 | 1 | 0 | 0 | 0 |
| **Average highs (°C)** | 30,8 | 29,0 | 27,4 | 25,3 | 20,9 | 18,0 | 17,0 | 17,3 | 22,1 | 24,5 | 27,5 | 28,6 |
| **Absolute Absolute (°C)** | 35,2 | 34,2 | 32,1 | 28,9 | 26,6 | 25,4 | 2,6 | 31,9 | 32,4 | 33,1 | 33,9 | 36,6 |
| **Rainfall (mm)** | 129,7 | 259,1 | 101,4 | 171,7 | 29,9 | 115,9 | 245,6 | 122,0 | 220,5 | 132,6 | 40,1 | 55,9 |
| **Rainy days** | 15 | 14 | 10 | 12 | 9 | 9 | 17 | 14 | 9 | 9 | 4 | 10 |

| | | | | | | | 2012 | | | | | |
|---|---|---|---|---|---|---|---|---|---|---|---|---|
| | **January** | **February** | **March** | **April** | **May** | **June** | **July** | **August** | **September** | **October** | **November** | **December** |
| **Average Temperature (°C)** | 23,9 | 25,0 | 21,5 | 18,0 | 16,3 | 12,8 | 11,4 | 17,1 | 16,9 | 20,0 | 22,1 | 24,1 |
| **Average minima (°C)** | 17,6 | 19,5 | 14,8 | 11,9 | 9,7 | 6,3 | 5,3 | 11,5 | 11,6 | 15,1 | 15,5 | 18,5 |
| **Absolute min. (°C)** | 12,6 | 13,6 | 3,0 | 3,7 | 1,9 | -5,1 | -2,3 | 5,6 | 4,2 | 9,5 | 11,0 | 5,6 |
| **Days below 10°C** | 0 | 0 | 0 | 0 | 2 | 7 | 14 | 1 | 10 | 0 | 0 | 0 |
| **Average highs (°C)** | 30,4 | 31,6 | 28,9 | 24,4 | 23,8 | 19,8 | 17,9 | 23,9 | 22,4 | 25,3 | 29,0 | 30,3 |
| **Absolute Absolute (°C)** | 36,3 | 37,6 | 37,2 | 32,9 | 30,2 | 29,7 | 29,9 | 31,2 | 31,1 | 31,7 | 33,5 | 38,6 |
| **Rainfall (mm)** | 42,9 | 136,7 | 112,0 | 75,4 | 66,5 | 21,5 | 219,3 | 67,4 | 67,0 | 169,8 | 51,7 | 214,4 |
| **Rainy days** | 7 | 12 | 9 | 6 | 5 | 7 | 9 | 7 | 11 | 13 | 5 | 7 |

continued APPENDIX 1. Average temperatures (°C), average minimum temperatures (°C), absolute minimum temperatures (°C), number of days with average temperatures below 10 °C, average maximum temperatures (°C), absolute maximum temperatures (°C), amount of rainfall (mm) and number of days with rainfall, monthly, from January 2011 to April 2014, at the UFRGS Agronomic Experimental Station, Eldorado do Sul, RS (30°06'16"S and 51°39'58"O).

| | **2013** | | | | | | | | | | | |
|---|---|---|---|---|---|---|---|---|---|---|---|---|
| | **January** | **February** | **March** | **April** | **May** | **June** | **July** | **August** | **September** | **October** | **November** | **December** |
| **Average Temperature (°C)** | 22,8 | 23,5 | 20,9 | 20,6 | 16,1 | 14,4 | 13,1 | 13,6 | 17,4 | 19,3 | 22,7 | 26,3 |
| **Average minima (°C)** | 16,8 | 18,4 | 15,9 | 14,9 | 10,6 | 9,8 | 7,4 | 8,7 | 12,2 | 14,2 | 14,6 | 20,8 |
| **Absolute min. (°C)** | 10,5 | 9,5 | 10,0 | 6,1 | 2,0 | 3,9 | -1,5 | 2,7 | 3,0 | 7,8 | 11,9 | 14,9 |
| **Days below 10°C** | 0 | 0 | 0 | 0 | 1 | 0 | 6 | 6 | 0 | 0 | 0 | 0 |
| **Average highs (°C)** | 29,1 | 29,7 | 26,1 | 26,3 | 21,6 | 19,1 | 18,9 | 18,6 | 22,6 | 24,4 | 27,9 | 31,8 |
| **Absolute Absolute (°C)** | 36,1 | 37,3 | 32,6 | 31,0 | 30,3 | 24,2 | 27,0 | 28,5 | 34,8 | 32,2 | 34,1 | 40,0 |
| **Rainfall (mm)** | 97,3 | 97,6 | 99,8 | 101,4 | 58,2 | 79,2 | 135,8 | 222,6 | 121,2 | 182,7 | 253,4 | 56,2 |
| **Rainy days** | 8 | 12 | 11 | 6 | 6 | 7 | 7 | 14 | 9 | 9 | 6 | 7 |

APPENDIX 2. Fruiting season of loquat, peach, araçá and caquize trees at the UFRGS Agronomic Experimental Station, Eldorado do Sul, RS (30°06'16"S and 51°39'58"O).

| **Fruit species** | **Jan** | **Feb** | **Sea** | **Apr** | **May** | **Jun** | **Jul** | **Aug** | **Set** | **Out** | **Nov** | **Dec** |
|---|---|---|---|---|---|---|---|---|---|---|---|---|
| Loquat trees | | | | | | | | **x** | **x** | **x** | | |
| Peach trees | | | | | | | | | **x** | **x** | **x** | **x** |
| Araçazeiros | | **x** | **x** | | | | | | | | | |
| Caquizeiros | | | **x** | **x** | | | | | | | | |

MIX
Papier aus verantwortungsvollen Quellen
Paper from responsible sources
FSC® C105338

Printed by Books on Demand GmbH, Norderstedt / Germany